T0296642

Spectrum of Belief

Transformations: Studies in the History of Science and Technology
Jed Buchwald, general editor

Myles Jackson, *Spectrum of Belief: Joseph von Fraunhofer and the Craft of Precision Optics*

Alan J. Rocke, *Nationalizing Science: Adolphe Wurtz and the Battle for French Chemistry*

Spectrum of Belief

Joseph von Fraunhofer and the Craft of Precision Optics

Myles W. Jackson

The MIT Press
Cambridge, Massachusetts
London, England

The MIT Press is pleased to keep this title available in print by manufacturing single copies, on demand, via digital printing technology.

Set in Sabon by The MIT Press.

Library of Congress Cataloging-in-Publication Data

Jackson, Myles. W.
Spectrum of Belief: Joseph von Fraunhofer and the craft of precision optics / Myles W. Jackson
p. cm.
Includes bibliographical references and index.
ISBN 978-0-262-10084-7 (hc. : alk. paper)—978-0-262-52723-1 (pb.)
1. Spectrum analysis—History. 2. Fraunhofer, Joseph von, 1787–1826—Knowledge—Optics. 3. Physicists—Germany—Biography. I. Title.
QC450.5.J33 2000
509.43'09'034—dc21 98-015177

to Mom and to the memory of Dad,
two true laborers

Contents

Acknowledgments

I have benefited greatly from the wisdom of many scholars over the six years I have worked on this project. In particular, Simon Schaffer taught me how to think like a historian, Peter Galison offered continued intellectual support as well as a year's funding from his Mellon Foundation Grant, and Jed Buchwald carefully read an earlier version of the manuscript and offered highly constructive comments and criticisms. Andrew Warwick also read an early draft and offered sound advice. His friendship over these past several years has been a source of inspiration. Others who have offered helpful comments along the way include David Cahan, Lorraine Daston, Michael Dettelbach, Menachem Fisch, Jeff Hughes, Rob Iliffe, Nicholas Jardine, Frank A. J. L. James, Michael Lynch, Iwan Morus, Kathy Olesko, Steven Shapin, and Ido Yavetz. And, of course, everyone who visited the tearoom of the department of history and philosophy of science at the University of Cambridge from 1987 to 1991 deserves thanks; it was a privilege to be part of that group. Jim Moore, who was in the group, offered me his home, hospitality, and friendship (not to mention his private library) during crucial summer archival trips back to Cambridge. Michael Becker made my stay in Berlin in 1991 to 1992 bearable.

I have offered versions of this material as seminars at universities, research institutes, and organizations. I thank all of those who attended and offered responses. A much earlier version of chapter 5 appeared as "Artisanal Knowledge and Experimental Natural Philosophers: Focusing on the British Response to Joseph von Fraunhofer's Optical Institute" (*Studies in History and Philosophy of Science* 25 (1994): 549–575). An earlier portion of chapter 6 appeared as "Illuminating the Opacity of Achromatic Lenses" in *The Architecture of Science*, edited by Peter Galison and Emily Thompson. The appendix and its figures appeared in "Buying the Dark Lines of the Solar Spectrum: Joseph von Fraunhofer and His Standard for Optical Glass Production" (*Archimedes* 1 (1996): 21–22) and are reprinted by permission of Kluwer Academic Publishers.

My research over the past six years has been funded by a Walther Rathenau Fellowship from the Volkswagen Stiftung, a DAAD (German Academic Exchange) Grant, a Deutsches Museum Research Grant, a Mellon Fellowship, a National Science Foundation Fellowship (grant SBE-9223098), a Travel Grant from the Morris Fishbein Center for the History of Science and Medicine at the University of Chicago, and a Herbert C. Pollock Award from the Dudley Observatory.

Finally, and most of all, thanks to Mom and Dad. Unfortunately, my father passed away during the writing of this manuscript. Both sacrificed much so that I would not have to toil as they had to throughout their lives.

Spectrum of Belief

1
Introduction

Most physicists and most historians of science are familiar with Joseph von Fraunhofer (1787–1826) and with the so-called Fraunhofer lines. Fraunhofer's "discovery"[1] of those dark lines that transect the solar spectrum and his subsequent work on the manufacture of optical glass and diffraction gratings either gave rise to or greatly advanced a large variety of new disciplines during the nineteenth century, including spectroscopy, x-ray technology, photochemistry, and (of course) stellar and planetary astronomy.

Fraunhofer was an apprentice-trained, working-class optical craftsman from a family with generations of experience in making and cutting glass. His father, Franz Xavier, and his paternal grandfather, Johann Michael, were master glassmakers from Straubing, Bavaria. On his mother's side, an impressive lineage of glassmakers stretched back to the beginning of the seventeenth century.

As was the case with many children from artisanal families of this period, Fraunhofer's youth was plagued by tragedy.[2] His mother had died in 1797, his father a year later. The orphaned Fraunhofer left Straubing in August of 1799 and moved in with his new master, the court mirror maker and decorative glass cutter Philipp Anton Weichselberger of Munich. Two years later, on the night of 21 July 1801, Weichselberger's house on Thiereckgäßchen collapsed. His wife was gravely injured, but the young apprentice was unscathed. Prince Elector Maximilian Joseph IV of Bavaria (later to become King Maximilian I) traveled to the scene of the disaster and was so moved by the young Fraunhofer that he later invited the youth to his castle in Nymphenburg and ordered his privy councilor, Joseph von Utzschneider, to see personally to the welfare of the child fate had spared.

Figure 1.1
Joseph von Fraunhofer (1787–1826). Portrait by R. Wimmer. *(Deutsches Museum, Munich)*

Figure 1.2
Fraunhofer's refractor of 1826, with an objective lens 9⅗ inches (9 Parisian inches) in diameter. With this telescope the Berlin astronomer J. G. Galle discovered the planet Neptune in 1846. *(Deutsches Museum, Munich)*

Figure 1.3
Prince Maximilian Joseph IV at the site of Weichselberger's collapsed house with the rescued Fraunhofer. *(Deutsches Museum, Munich)*

The Prince Elector provided the child with a gift of 8 Karolinen[3] and instructed the bewildered boy to contact him should he need further support.

After continuing his apprenticeship with Weichselberger and working with the Benedictine-trained optician Joseph Niggl on Sundays from 1801 to 1804, Fraunhofer became a journeyman of decorative glassmaking and cutting in 1804. But he soon became bored with his work and angered by the tyrannical behavior of Weichselberger, who did not permit Fraunhofer to use a reading lamp at night to study optical theory and optical glass practice. In addition, Weichselberger attempted to prevent him from regularly attending the Feiertagsschule, a school specifically designed to educate working-class boys. Fraunhofer sought the advice of the well-connected Utzschneider. He explained his situation to the by-then-former privy councilor, and Utzschneider responded by offering the then-nineteen-year-old journeyman a position in his Optisches Institut (Optical Institute) assisting in the manufacture of achromatic lenses for refracting telescopes and ordnance surveying instruments. With the aforementioned financial gift from the prince elector, Fraunhofer was able to buy out his contractual agreement with Weichselberger (for about 50 florins) and to purchase a glass-cutting machine capable of handling optical glass.

In 1806 Fraunhofer embarked upon a career as an optician. Within 15 years, the orphaned lad from the most humble of origins produced the finest achromatic lenses and prisms the optical community had ever seen.

The Fraunhofer story—a rags-to-riches story worthy of a Hollywood production—has been the object of much hagiographic history writing. One portion of the Fraunhofer literature celebrates him as the "father of spectroscopy" and the "discoverer of the emission and absorption lines."[4] Another portrays him as the "father of German optics" in an attempt to argue for German hegemony in both the physical sciences and the optical industry.[5] Much of this secondary literature, however, has ignored the crucial role that Fraunhofer's manual skills played in the development of nineteenth-century physics.

Although Fraunhofer's life was indeed fascinating, what follows is not a biography. Rather, I will use Fraunhofer's career as a heuristic tool to probe the relationships between science and society in Bavaria and Great Britain early in the nineteenth century. Fraunhofer lived during a historical period that witnessed critical exchanges between artisans (such as opticians and scientific instrument makers) and experimental natural philosophers. On the one hand, the experimental natural philosophers needed the skilled artisans, as they rarely produced their own instruments. Industrialization gave rise to a precision technology that required a specialization that most experimental natural philosophers did not possess. On the other hand, artisanal knowledge was thought by natural philosophers to be tainted. First, people who performed manual labor, particularly in industrialized Britain, were regarded as the inferiors to intellectual savants by the late eighteenth century.[6] Second, skilled artisans did not belong to the so-called Republic of Letters, as they were generally not university educated, and their craft knowledge was a tightly guarded guild secret.[7] Hence, artisans, although essential to the scientific enterprise, were often not granted the status enjoyed by experimental natural philosophers. This book explores the relationship between artisanal knowledge and experimental natural philosophy by first analyzing Fraunhofer's contributions to geometrical and technical optics and then examining the responses to his work by various German and British experimental natural philosophers.

Chapter 2 offers a brief history of the state of achromatic-lens production before Fraunhofer's work. Two general sorts of optical investigations were pursued during the eighteenth century and the first two decades of the nineteenth. One was a trial-and-error method of tinkering used by skilled artisans in the manufacture of optical lenses and instruments, typified by Chester Moor Hall and John and Peter Dollond. The second sort of optical investigation was that of the savant, as epitomized by Isaac Newton, Leonhard Euler, Samuel Klingenstierna, A. C. Clairaut, Jean Le Rond d'Alembert, and John Herschel. By the early nineteenth century, savants such as Herschel and David Brewster attempted to bridge the ever-widening chasm between natural philosophers and opticians by offering articles with simplified mathematics so that craftsmen could produce better lenses.

Chapter 3 reconsiders Fraunhofer's history by placing his optical glass production in the contexts of the reformed politics of Bavaria during Napoleon's occupation, the growth of ordnance surveying, and the secularization of Benedictine monasteries. The need for precision optical and ordnance surveying instruments was felt throughout the German territories by astronomers and ordnance surveyors who were producing more accurate maps of their regions. Napoleon needed topographical maps of the new territories his troops had acquired, and regional rulers such as King Maximilian I based economic reforms on property taxes; hence, property boundaries had to be determined with a much greater accuracy. Until 1800, Great Britain had been the purveyors of optical lenses to the Continent. But the blockade of 1806 cutting off Great Britain from the Continent prevented the British from exporting their optical lenses. Thus, not only did Napoleon assist in the destruction of Britain's optical empire; he also guaranteed a market for Fraunhofer's products throughout the Continent. This temporary shift in optical monopoly had profound ramifications for both Britain and Bavaria. For Bavaria, it signaled the transformation of a rural agricultural state into a leader in technological and scientific advances. For Britain, Fraunhofer's usurpation of their optical industry provided yet another example of the continued decline of British science, as lamented by Charles Babbage and David Brewster. Much was at stake.

It was in this context that Fraunhofer conducted his famous research on achromatic lenses and prisms using the solar spectral lines. Fraunhofer drew upon the secretive monastic labor of Benedictine monks, who were generally well versed in optical theory, and the secretive guild practices of artisanal craftsmen from the area of Benediktbeuern, Bavaria. Fraunhofer's laboratory was a secularized Benedictine monastery whose architecture reflected the importance of labor, silence, and secrecy to the Benedictine order. Fraunhofer and Utzschneider had many resources at their disposal in order to ensure that the recipe for and the skilled manipulations of the Optical Institute's achromatic glass remained closely kept secrets.

This book as a whole illustrates how the boundaries of economics and science, which are often taken for granted, were, during this period, blurred. Fraunhofer's artisanal skills, combined with an aggressive, reform-minded Bavarian government, gave rise to a precision optical technology that was to become the envy of the world. Freiherr Maximilian von Montgelas's governmental reforms, backed by King Maximilian I, were intended to launch Bavaria to the technological, scientific, and economic levels of post-Revolutionary France. These reforms shifted the bureaucratic power from the clergy of the Roman Catholic Church to a new group of men: the Beamten, or civil servants. Britain, on the other hand, taxed its glassmakers, while the Industrial Revolution lowered the social status of skilled artisans. The economic conditions of early-nineteenth-century Britain inevitably led to that nation's temporary downfall in optical technology in the first half of the nineteenth century. What skills were necessary for the improvement of both experimental natural philosophy and society, and who possessed such skills, were major questions for Britons and Germans during the first half of the nineteenth century.

How some German investigators of nature responded to Fraunhofer's artisanal knowledge is the subject of chapter 4. Although all were convinced of the merits of Fraunhofer's optical lenses and prisms, debates erupted as to whether he was a Naturforscher or simply a gifted artisan. Several investigators of nature objected to his practice of secrecy. They argued that, since guild knowledge was secretive, it was anathema to the openness of scientific knowledge. They also protested when Fraunhofer, who had never attended a university, was allowed to join their ranks in the Royal Academy of Sciences in Munich.

Chapters 5 and 6 detail the British response to Fraunhofer. Precisely because Fraunhofer's method and recipes for optical glass production were closely guarded secrets, whether or not such knowledge could be reconstructed without witnessing the actual manual, skilled labor became a critical question in Britain during the 1820s and the 1830s. British experimental natural philosophers and opticians, with the financial support of the Royal Society of London and the Board of Longitude, attempted to recapture the world's market for achromatic lenses.

Michael Faraday attempted to reverse engineer (to use an anachronistic term) Fraunhoferian lenses and prisms by analyzing their chemical compositions and constructing glass blanks with the same ingredients in their proper ratios. Faraday offers an interesting comparison to Fraunhofer. He, too, came from a humble household, and his claim to fame was also his experimental brilliance. His trial-and-error practices of optical glass manufacture resembled Fraunhofer's, although he clearly was not as successful. But crucial differences exist between the two. Fraunhofer had busied himself with glass his entire life. Working with glass was his family tradition, and the manufacture of optical lenses and prisms was his life. Faraday, however, was commissioned to work on optical glass. He saw it as a hindrance to his other experimental work, which he viewed as much more important than tinkering with glass. But the most crucial distinction discussed in this work was Fraunhofer's reckoning that the artisanal practice of optical glass had to be created from the knowledge deeply embedded in the craftsman's culture and hands. Faraday, quite to the contrary, originally believed that this artisanal knowledge could be read and created from the analysis of the final products. George Dollond, Britain's foremost optician (grandnephew of John), ground and polished Faraday's glass blanks into lenses, and John Herschel, Britain's leading experimental natural philosopher of the period, determined the refractive and dispersive indices of those lenses.

The British efforts proved to be futile. Indeed, the British response to Fraunhofer's artisanal knowledge is very informative to cultural historians of science. Brewster, Faraday, and Herschel all debated whether Fraunhofer's artisanal knowledge could be communicated—and, if so, how. They held differing, and often antithetical, views on the relationship between "scientific" and "artisanal" knowledge.

Chapter 7 details the end of the Optical Institute's monopoly after Fraunhofer's death. Fraunhofer's secrets of manufacture accompanied him to the grave. His artisanal knowledge was such that, after his death, even the apprentices who had worked most closely with him, in the same glass hut and with the same equipment, achieved only limited success in the manufacture of optical glass. In the ensuing decades, Germans once again needed to import optical glass and instruments from Britain and France.

After German unification, Fraunhofer became a powerful historical example of the benefits of merging scientific research and technological innovation with industrial and state support. As it turns out, our knowledge of Fraunhofer's place in history has been greatly informed by numerous historical accounts offered during the Kaiserreich. Hermann von Helmholtz and Ernst Abbe used Fraunhofer to bolster their attempts to unify the efforts of science, technology, and industry during the 1870s and the 1880s. Indeed, precisely because optical hegemony had left German soil, Helmholtz and Abbe pleaded with the Prussian government to subsidize Abbe and Otto Schott's optical glass experiments at the Carl Zeiss Works in Jena during the 1880s.

Chapter 8 discusses how Fraunhofer has been depicted in the history of physics. During the late nineteenth century, German scientists cleverly utilized the genre of history in general, and the persona of Fraunhofer in particular, to explain and extol their nation's preeminence in the physical sciences. They, in Eric Hobsbawm's words, invented a tradition.[8]

In one respect, my story is an attack on those who claim that universal replication is proof that contextual studies of science merely offer a window dressing to the more important epistemological character of science. It was, after all, Hans Reichenbach and Karl Popper who emphasized the distinction between the context of discovery or invention and the context of justification or replication.[9] Indeed, logical positivists continually referred to this distinction. Although both philosophers claimed that "social factors" could influence a discovery, they insisted that independent confirmation (i.e., replication) of that discovery should be free from social factors; it was "disinterested" and "objective." Philosophers of science delved into justification, confirmation, or falsifiability because, as they claimed, it was the stuff of "real" science:

> Only when certain events recur in accordance with rules or regularities, as is the case of repeatable experiments, can our observations be tested—in principle—by anyone. We do not take even our own observations quite seriously, or accept them as scientific observations, until we have repeated and tested them. Only by such repetitions can we convince ourselves that we are not dealing with a mere isolated

"coincidence," but with events which, on account of their regularity and reproducibility, are in principle inter-subjectively testable. . . . Any empirical scientific statement can be presented in such a way that anyone who has learned the relevant technique can test it.[10]

But Thomas Kuhn pointed out that "social factors" are just as important to the context of justification as they are to the context of discovery.[11] The importance of skills and practices in the replication of scientific knowledge has been a central topic of interest in the history, sociology, and philosophy of science in recent years. Philosophical studies by Mary Hesse, Paul Feyerabend, Thomas Kuhn, and N. R. Hanson and historical studies by Gerald Holton began the assault on the doctrine of logical postivism, which exclusively investigated "scientific discoveries" and their justification.[12] More recently, sociologists have argued that replication is a much more active process than philosophers have previously suggested, and that it is predicated upon recreating a portion of the laboratory in some other place, be it in another laboratory or out in the wild.[13] Several sociologists of science have introduced the notions of the "skill-ladenness" and "practice-ladenness" of the scientific enterprise to science studies.[14] Other sociologists, historians, and philosophers have argued that replication is accomplished only by the personal witnessing of the practices of someone who has produced the desired artifact or knowledge. That is to say that successful replication is predicated upon the transmission and acquisition of material culture.

The themes of skills and practices tie this work together at the most fundamental level. The importance of experimental skills and practices to the scientific enterprise has been the subject of several recent studies.[15] A small portion of this sociological literature (in particular, the work of Harry Collins) has drawn upon the work of Michael Polanyi and Ludwig Wittgenstein. Of particular interest are the many ways in which an individual follows a set of rules and practices.[16] In *Personal Knowledge*, Polanyi states that skillful performance is

an art which cannot be specified in detail [and therefore] cannot be transmitted by prescription, since no prescription for it exists. It can be passed on only by example from master to apprentice. This restricts the range of diffusion to that of personal contacts, and we find accordingly that craftsmanship tends to survive in closely circumscribed local traditions. . . . Again, while *the articulate contents of science* are successfully taught all over the world in hundreds of universities, *the unspecifiable art of scientific research* has not yet penetrated to many of these.[17]

Polanyi's model, quite clearly, was one of empowerment, in two different yet related ways. First, the master possesses all of the power in respect to his or her apprentice. Polanyi strongly implies that knowledge (and hence power) flows in one direction only: from teacher to pupil. Second, once the apprentice has obtained the necessary tacit knowledge, he or she is empowered with respect to those in a community who do not possess such knowledge. Polanyi's scheme is one of extreme elitism. As he admitted,

> To learn by example is to submit to authority. You follow your master because you trust his manner of doing things even when you cannot analyze and account in detail for its effectiveness. By watching the master and emulating his efforts in the presence of his example, the apprentice unconsciously picks up the rules of art, including those which are not explicitly known to the master himself. These hidden rules can be assimilated only by a person who surrenders himself to that extent uncritically to the imitation of another. A society which wants to preserve a fund of personal knowledge must submit to tradition.[18]

Although this model of skill transmission is useful in some cases, rarely are apprentices submissive, empty slates upon which a dominating master writes his wisdom. Jerome Ravetz elaborates upon Polanyi's *Personal Knowledge* in his work *Scientific Knowledge and Its Social Problems.* Ravetz argues that science is quintessentially "craftsman's work." He emphasizes the point that the mastery of scientific knowledge depends in part upon traditional craft knowledge, communicated by "precept and tradition."[19]

Quite famously, Collins applied Polanyi's notions of tacit knowledge to the building of the TEA laser. Collins's rationale was to argue for an enculturation account of learning skills—rather than an algorithmic model—in what had previously seemed to be the science that was universal, and free from culture: physics. If physicists did not use an algorithmic model of learning, but an enculturated model, then cultural histories of physics could be written, and physics could now possess a history, or so Collins reckoned. No longer did historians of science need to follow the cue of philosophers of science. Collins argued that, in the attempt to replicate the construction of the TEA laser,

> the flow of knowledge was such that, first, it travelled only where there was personal contact with an accomplished practitioner; second, its passage was invisible so that scientists did not know whether they had the relevant expertise to build a laser until they tried it; and third, it was so capricious that similar relationships between teacher and learner might or might not result in the transfer of knowledge.[20]

By applying Wittgenstein's and Polanyi's enculturation model of skill acquisition to physics, Collins's work has, of course, given rise to many fruitful analyses. But surely individuals can acquire many kinds of skills by simply following an algorithmic model. It obviously depends on the degree of complexity of those skills. A more recent study by Pinch, Collins, and Carbone has conceded that "the enculturation model has been a success, but it cannot be the whole story."[21] Secondary studies that suggest that skills and practices and their transmission and communicability are the properties of communities ignore *how* information can be explicitly transmitted and acquired as part of the process of learning a skill or practice. The more recent work of Pinch, Collins, and Carbone introduces a notion of "second-order studies of skill" which accepts that skill and practice acquisition occurs within a particular cultural context, but their work continues beyond that point in order to examine which aspects of skills can be explicated and which cannot. Kathryn Olesko's work has demonstrated that fewer skills are tacit—and many more are expressible—than Polanyi or Collins perhaps might realize. The very process informing the basis of a research school is the transmission of craft skills from colleague to colleague or from professor to student.[22]

My point is this: not all scientific procedures can be written down algorithmically. That is why, for example, postdoctoral laboratory bench training is so crucial to the development of a scientist's career. As Ravetz suggests, "Every field [of science] has its own special tools, of which some require an extensive technical knowledge, supplemented by craft experience, for their use. It is for this reason that a lengthy apprenticeship is necessary before anyone can embark on independent work in a developed scientific discipline."[23] It is important to stress at the outset that my story is not another sociological study of tacit knowledge. One must differentiate between *knowledge that was never communicated* and *knowledge that is by nature incommunicable.* 'Tacit' is rather ambiguous. It can either mean a skilled activity with a hidden transfer of knowledge that is not articulable, or simply unarticulated knowledge taught by precept and example. The former meaning is particularly applicable to material activity, whereas the latter meaning is very relevant to natural philosophy or mathematics.[24] As I argue in this work, experimental natural philosophers such as Herschel certainly appreciate the importance of training by example, but are puzzled by problems posed when there is no one to articulate a certain type of knowledge. Contrary to what some studies wish to suggest, secrecy is not inherently an example of tacit knowledge. Although the technique to reproduce the dark lines was indeed communicated by Fraunhofer, information regarding the manufacture of his optical lenses and prisms was not. Hence, whether or not someone could replicate these objects if given enough time, financial support, willingness, and the recipes is actually a moot point. Historically, we can never hope to uncover this information. Historians need not simply abandon the project, however; instead, they might ask *how historical actors claim that artisanal knowledge can, in principle, be communicated.* By analyzing what my actors claimed are skills and how such skilled knowledge could be communicated, I began to piece together certain cultural activities that had previously seemed to be unrelated—namely, the Bavarian and British responses to Fraunhofer and the role of secrecy, the reform of British patent laws, the role of the Mechanics' Institutes in the stimulation of British manufacturing in the 1820s and the 1830s, British patronage of science, mechanization, and the decline-of-science debate. What skill was, how it could help society at large, and who possessed it were highly contentious issues in Britain and Bavaria during the period being considered here, when physics was undergoing rapid change. Therefore my story offers insight into the relationship between science and society. Since both physics and society were being reconstructed during this period, the reader can see just how culturally embedded the scientific enterprise was. Topics dealing with skills and practices in physics were also simultaneously topics of concern for reformed Britain and Bavaria, and the politics of labor can elucidate those topics. Because sociologists of science have concentrated much of their work on the twentieth century, they have often been guilty of accepting prima facie that the established structure of science and its relationship with a broader society have always been as they find them now. Throughout this book, I will attempt to historicize the sociologists' notions of skills and practices, which, after all, are historically located: they are spatially and temporally specific.

Over the past 15 years, there has been a growing literature on the importance of artisans to the rise of science during the seventeenth, eighteenth, and nineteenth centuries. Elaborating on the earlier classic work of Edgar Zilsel and the role of artisans during the Scientific Revolution, Jim Bennett has convincingly demonstrated the importance of seventeenth-century instrument makers such as Robert Hooke to the development of the mechanical philosophy.[25] Steven Shapin's model of "invisible technicians" has served as a model for the construction of artisanal histories.[26] Anne Secord has argued for the role of artisans in the rise of botanical studies in England during the eighteenth century.[27] H. Otto Sibum has given us a study of the importance of the skilled labor found in England's brewing industry of the 1830s to James Joule's research on the mechanical equivalent of heat.[28] Simon Schaffer has underscored the importance of the skills of dyers in early-eighteenth-century England to Stephen Gray's construction of an "electric planetarium," and Iwan Morus has discussed the relationship among artisans, natural philosophers interested in electricity, and machines in early-nineteenth-century London. As Schaffer rightly points out, relations between artisans and gentry point to tensions over the status of manual labor in experimentation.[29]

This book explores four traditions of the nineteenth-century understanding of what made optical precision-technological practice possible.[30] The first tradition pertains to secrecy, in which the key to practice is embodied in arcane knowledge to be preserved by an artisanal elite—a tradition dating back to medieval guilds. Such knowledge must have been at least in part transmissible as direct knowledge; hence it was tightly guarded. The second relevant theme—which is related to, yet distinct from, the first—is the craft-skill tradition, in which practice is socially embodied in the knowledge of artisans and is passed on through apprenticeship. Such a tradition was rather famously discussed by the Encyclopedists of Enlightenment France. By the craft-skill tradition I mean the amassed knowledge of the skilled manipulations typical of artisans: their trial-and-error techniques of producing artifacts, such as scientific instruments. Much of this type of knowledge was passed down from generation to generation, not in written form, but rather by word of mouth and demonstration. Not all artisanal knowledge is tacit, and not all tacit knowledge is necessarily artisanal. Also, the artisanal knowledge discussed in this book is fundamentally a possession of a particular social class. The way in which artisans managed their labor and conducted their work was intricately linked to the structure of their guilds. To ignore this socio-economic dimension of artisanal knowledge is to miss a major point. The rational or scientific tradition is the third relevant theme. Here, practice is embodied in scientific knowledge, in theory, or in the public rules of action. The fourth and final theme is the managerial one, in which practice is embodied in the entrepreneurial skills necessary to organize and coordinate that practice, ensuring its viability in future generations. How contemporaries combined these traditions, emphasized some over others, or simply excluded some of them in their various kinds of involvement with precision optics shapes the narrative of this book. A study of Fraunhofer can be used to explore the topography for a historical investigation of these factors as the basis for precision-technological practice in nineteenth-century optics.

2

Optics before Fraunhofer

Light became [the French's] favorite subject on account of its mathematical obedience and freedom of movement. They were more interested in the play of its colors, and thus they named after it their great enterprise, the Enlightenment.

—Novalis, *Christendom or Europe*

Such skill [of a genius] cannot be communicated, but requires to be bestowed directly from the hand of nature upon each individual.

—Immanuel Kant, *The Critique of Judgement*

In order to appreciate Fraunhofer's contribution to optics,[1] we need to turn away from the German territories and focus on eighteenth-century England and France. Although Leibniz and Euler contributed greatly to optical science, their works were generally written in either Latin or French. By the late eighteenth century, optical texts appearing in German were generally manuals read by opticians and were often viewed with contempt by many experimental natural philosophers. This chapter discusses two traditions of optics: the philosophical writings of natural philosophers and the texts detailing the trial-and-error procedures of instrument makers and opticians.

In his *Opticks* (1704, book 1, part 2, proposition 3, problem 1, experiments 7 and 8), Isaac Newton tested whether chromatic aberration in telescopes could be corrected by combining two lenses of differing refractive indices. The chromatic aberration of one lens would have to cancel out the other, reproducing white light, but this (Newton asserted on experimental grounds) would occur only if the emergent light was parallel to the incident light, rendering lenses utterly useless. Hence, quite famously, Newton abandoned the refracting telescope for the reflecting telescope, which employed mirrors.

Newton's erroneous pronouncement that net refraction is always accompanied by dispersion greatly impeded research on achromatic lenses for three decades thereafter. Between 1729 and 1733, a London barrister named Chester Moor Hall discovered that one could partially overcome chromatic aberration by combining glasses of opposite powers: a concave lens made of flint glass and a convex lens made of crown glass[2] (figure 2.1). Such a doublet corrected chromatic aberration for the red and violet rays. The London optician George Bass constructed and sold the actual lenses according to Hall's specifications.

In 1747 the German mathematician and natural philosopher Leonhard Euler published a memoir in the proceedings of the Berlin Academy of Sciences in which he argued that, contrary to Newton's claims, it was indeed theoretically possible to correct for both spherical and chromatic aberration in object lenses. Using the physiology of the eye as his model, Euler argued that water placed between two concave lenses should correct chromatic aberration. Instead of directly observing the refractive and dispersive powers of the glass and water, however, he attempted to derive mathematically a general law for achromatic telescopes.

London's leading optician of the period, John Dollond, was not impressed by Euler's memoir. In a letter addressed to the astronomer James Short and published in the *Philosophical Transactions of the Royal Society*, Dollond originally defended Newton's position against Euler's theorem. He criticized a hypothesis of Euler's, which the Englishman claimed was

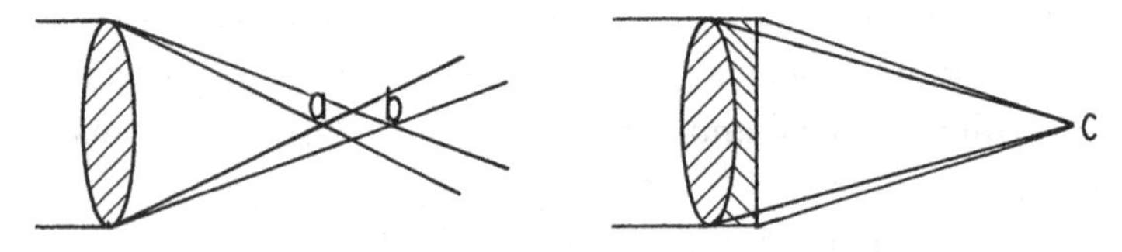

Figure 2.1
Chromatic aberration occurs when white light passes through two media with different densities, such as air and a glass lens. White light is divided into its components, the colors of the spectrum. Each color has a unique index of refraction, violet possessing the greatest index and red the least. a: the focal point of the violet rays. b: the focal point of the red rays. Chromatic aberration can be corrected by juxtaposing a double-concave flint lens with a plano-convex crown lens. This lens combination focuses the red and violet rays to a point (c). It would take the combined effort of Otto Schott and Ernst Abbe in the 1880s to correct the red, violet, and yellow rays via their apochromatic lenses.

"destitute of support either from reason or experiment, though it be laid down as the foundations of the whole fabrick."[3] Dollond argued against Euler's assumption that the ratio of refraction out of air into glass of the mean refrangible rays is the same power as for the least refrangible and that the ratio of refraction out of air into water of the mean refrangible rays is the same power as the least refrangible.[4] Dollond concluded his piece by declaring that "Mr. Euler's theorem is intirely founded upon a new law of refraction of his own, but that, according to the laws discovered by experiment, the aberration arising from the different refrangibility of light at the object-glass cannot be corrected by any number of refractions whatsoever."[5]

Euler's memoir of 1747 also aroused the interest of the Uppsala mathematician Samuel Klingenstierna, who decided to recreate Newton's experiments. He examined the path of light rays through a combination of glass prisms and concluded that Newton's results were only applicable to prisms of small apex angles.[6] In 1754 Klingenstierna published a mathematical demonstration proving that Newton's assertion of the impossibility of constructing an achromatic refracting telescope was fallacious. In 1755 Klingenstierna informed Dollond of his work, and Dollond immediately began working on the problem of building an achromatic telescope by means of experimentation, after having originally defended Newton's results against Euler's theorem.

Dollond, much like Newton more than 50 years earlier, cemented two plates of parallel glass at their edges forming a prismatic vessel, sealed the bases, and filled it with water. He then placed a glass prism with one of its edges turned upward inside the larger prismatic vessel,[7] so that a ray of light was transmitted through both of these refracting media (the water and the glass). Dollond proceeded to decrease or increase the angle between the two glass plates until the two refractions were equal, which was precisely the point at which an object being viewed through this double prism appeared as it looked to the naked eye (i.e., neither depressed nor elongated via refraction). According to Newton, the object should have appeared to be colorless through the double prism. Dollond was surprised that this was far from the case. Indeed, his experiment showed that "the divergency of light by the prism to be almost double of that by the water. For the object, tho' not all that refracted, was yet so much infected with prismatic colour."[8] He reckoned that if the refracting angle of the water vessel could be increased, the divergence of the colored rays (or dispersive power) would be greatly diminished if not totally eradicated.

Dollond then proceeded to perform an experiment with prisms possessing smaller angles. He ground a prism of common plate glass to an angle of slightly less than 9°, and then made a wedge-like vessel filled with water similar to the one described above so it refracted equally with the glass prism.[9] Objects viewed through this double prism appeared in their normal shape (neither elongated nor depressed). Although the refractions were equal, the dispersive power of the glass was far greater than that of the water, and the objects viewed were very much discolored. He increased the angle of the water prism until the objects appeared free of color. At that point the refraction of the water was 5/4 the refraction of the glass.

The next step was to "grind wedges of different kinds of glass, and apply them together, so that the refractions might be made in contrary directions, in order to discover, as in the foregoing experiments, whether the refraction and divergency of the colours would vanish altogether."[10] This Dollond did in late 1757. He soon discovered the "refractive qualities" of different types of glass: Venice glass had a lower "refractive quality" than English crown glass, which was still less than that of English flint glass.[11] He ground a prism of white flint glass to 25° and another prism made of crown glass to 29°; their refractive powers were nearly the same, but their dispersive powers were rather different. He ground several more prisms of crown glass with different angles until one of them possessed the same dispersive power as the flint glass. When he placed those prisms together, the object viewed was totally free of color. When measuring the refractive power of each prism, he found that the ratio of the refractive power of flint glass to crown glass was nearly 2 to 3. Lenses could then be cut from these two prisms: the crown glass was cut into a convex lens, and flint glass into a concave lens. And, as the refractive powers of the two lenses are the inverse of the ratio of their focal lengths, the focal lengths of two lenses are the inverse of the ratio of the refractive powers of the two prisms from which they were cut, or 3 to 2.[12] Dollond then focused his attention on spherical aberration and noticed that by means of a series of grindings of the lenses' surfaces he could rid the lenses of such defects. Dollond produced superior achromatic telescopes, which he supplied to countries throughout Europe. He was awarded with Britain's highest scientific honor, the Copley Medal, and with membership in the Royal Society of London.

After reading Dollond's piece, Klingenstierna began to tackle the problem of achromatic and aplanatic lenses (i.e., lenses free of spherical aberration) by using geometrical optics. His research culminated in an impressive mathematical treatise in 1760, a slightly modified version of which appeared in 1762 in the *Proceedings of the St. Petersburg Academy of Sciences* and won him the prize for the "best investigation for the possibility of eliminating the imperfections of the telescope and microscope brought about by means of the differing refrangibility of light and spherical aberration."[13] A Latin translation of his 1760 Swedish essay was published in the *Philosophical Transactions* in 1761,[14] and the French savant A. C. Clairaut translated it for the French Academy.

Beginning in 1761, Clairaut replicated Dollond's experiments and noticed that objects seen through supposedly achromatic lenses were in fact surrounded by green and wine-colored fringes resulting from the fact that the dispersive powers of the two glasses were not perfectly matched at all colors.[15] Clairaut also reckoned, correctly as it turns out, that crown and flint glass have different partial dispersions, and he demonstrated that the image generated by a two-lens combination must necessarily be accompanied by the secondary spectrum.[16] He determined the refractive and dispersive indices of several types of glass by means of a device, composed of a glass prism with a variable top angle that permitted light rays to pass through a plano-convex lens at various points between the lens's rim and its center.[17] He also deduced theorems for pairing the glasses in order to reduce the effect of the secondary spectrum. But he could never totally eradicate the secondary spectrum, as he never had a technique for determining the refractive indices for precise portions of the spectrum.

Clairaut at first limited his optical investigations to a double objective composed of crown and flint glass. Owing to the severe curvature of the surfaces of the lenses and the aberration arising beyond the optic axis, he concluded that flint glass should not be used with crown glass to correct for chromatic and spherical aberration.[18] He decided instead to try a different type of glass whose diffractive power was greater than that of English flint glass. However, this type of glass, called strauss, contained too much lead oxide to be useful for achromatic lenses; striae and veins, due to the density of lead oxide, often rendered the glass unusable.[19]

Clairaut's wish was to free France from its reliance on the flint glass that England exported, which was required by law to be inferior to the flint glass supplied to British opticians. He offered his mathematical calculations to two Parisian opticians, de L'Estang and Antheaulme, with the hope that their craftsmanship, combined with his theory and geometrical optics, could produce achromatic lenses superior to those of John Dollond and his son Peter.[20] Clairaut also discussed the problem of three types of extra-axial aberrations in double and triple object glasses: astigmatism, curvature, and coma.[21] He realized that even when chromatic and spherical aberrations were corrected, parallel rays incident on the objective obliquely to the optic axis intersected outside the focal plane, giving rise to these forms of aberrations. Although he successfully reduced astigmatism and curvature to reasonable limits, coma was still a problem.

Back in England in the early 1760s, Peter Dollond was working on triple lens constructions that could correct for chromatic aberration better than his father's. Peter's third lens both reduced the secondary spectrum and offered an improved correction for spherical aberration.[22] Between 1764 and 1768, Jean Le Rond d'Alembert published his work on achromatic lenses in volumes III and IV of his *Opsuscules Mathématiques*. In this treatise, d'Alembert differentiated longitudinal aberration (caused by the fact that different colors have different foci along the optic axis) from lateral or transverse aberration (caused by the fact that different colors possess different magnifications, which affects telescope eyepieces).[23]

In the 1760s, R. J. Boscovich, a Jesuit professor at Padua, was trying to design object lenses free of both chromatic and spherical aberration and eyepieces free of chromatic effects.[24] He reported the development of an instrument, similar to Clairaut's, for studying dispersions.[25] It was composed of a water prism with a variable top angle, which could be measured on a graduated arc, and a glass prism. A short spectrum was thereby cast on a screen, and Boscovich altered the angle of the water prism until the secondary spectrum was reduced considerably.[26] In 1773, after various improvements of the device (which he called the vitrometer), Boscovich concluded that Newton was incorrect in assuming proportionality between refractive and dispersive powers in different substances, and that Clairaut was correct in arguing that the relative positions of the spectral colors are not always the same.[27] However, he did not have sufficient information to calculate the radii for the surfaces of a system of telescopic lenses of flint and crown glass free of the secondary spectrum, since he did not know how to measure the refractive indices for each colored ray.[28] He did, however, demonstrate a method of uniting three spectral colors by means of three media, each possessing different refractive and dispersive powers.[29]

The Edinburgh professor of astronomy, Robert Blair, also researched the elimination of the secondary spectrum. In 1791 he read two papers to the Royal Society of Edinburgh (one on 3 January and one on 4 April) detailing alternative methods of constructing achromatic telescopes.[30] He researched the dispersive qualities of many substances in addition to the typical crown and flint glasses, "in hopes of finding some dispersive medium, which would separate the differently refrangible rays in the same proportion in which crown-glass does, and thus afford a method of refracting all of them alike, and consequently without colour."[31] Failing to achieve this, he decided to combine two oils with very different dispersive powers, both greater than that of crown glass. The less dispersive oil was used as a convex lens; the oil with the greater dispersive power was made into a concave lens.[32] The secondary spectrum (a term Blair coined) of green and purple fringes still appeared, but its width was greatly reduced.[33] From these trials Blair was able to conclude

> that if I took an achromatic convex lens, composed of the two essential oils, and combined it with an achromatic concave lens of a longer focal distance, composed of crown-glass and either of the essential oils, I should be able, through such a double compound object-glass, to converge the rays to a focus, without any aberration whatever from the difference of refrangibility of light. For if the compound convex and compound concave are properly proportional to each other, the secondary spectrums, or fringes of green and purple, may be rendered of the same breadth in both lenses.[34]

Blair never completely succeeded in eliminating the secondary spectrum, although he did make some progress. The combination of crown glass with minute bits of metal dissolved in muriatic acid (now called hydrochloric acid) noticeably shortened the secondary spectrum. Blair's work was the origin of the technology of liquid-lens telescopes, which David Brewster, Archibald Blair (Robert's son), and Peter Barlow attempted to resurrect in the 1820s in response to Fraunhofer's superior achromatic telescopes.

Although geometrical optics was relatively advanced, the practice of making achromatic lenses certainly was not. One of the most serious problems facing opticians during the second half of the eighteenth century and the early years of the nineteenth was the quality of crown and (especially) flint glass. This was particularly troublesome for the British opticians (including John, Peter, and George Dollond), inasmuch as they did not manufacture optical glass themselves; instead, they purchased glass blanks (rectangular chunks or disks of optical glass) produced by glassmakers in London, which they then ground and polished. Glassmakers who could manufacture optical disks larger than 4 inches in diameter were highly sought after. Larger disks often suffered from such imperfections as bubbles or striae, rendering the glass unusable for optical purposes.

The most widely used text for eighteenth-century English opticians trying to cope with the limitations of the glassmakers was Robert Smith's *A Compleat System of Opticks in Four Books, viz. a Popular, a Mathematical, a Mechanical, and a Philosophical Treatise,* published in 1738. Smith was Master of Trinity College, Cambridge, Professor of Astronomy and Experimental Philosophy, and Master of Mechanics to His Majesty. Book 3 of Smith's *Opticks*, a mechanical treatise, detailed Christiaan Huygens's methods of grinding and polishing lenses for refracting and reflecting telescopes. Smith claimed that, since "Mr. Huygens's treatise is esteemed the best of any yet extant, I have taken care that nothing material therein contained should pass unobserved."[35] Other chapters in the mechanical treatise included "How to Center an Object Glass" and "Methods of Casting, Grinding, and Polishing Methods for Reflecting Telescopes."[36]

Smith's work was sufficiently influential during the late eighteenth century to warrant an abridged edition for university students in 1778.[37] Interestingly, the mechanical treatise of the original was totally omitted: included were only several chapters from the theory of light and colors (book 1) and the mathematical portion of book 2. This abridged work hints at the extreme sundering of optical theory from practice that had emerged during the early eighteenth century, but had become extreme in Britain by century's end, with students of natural philosophy concentrating exclusively on physical and geometrical optics, and opticians on technical optics, limited to basic algebra and the rule of sums, relevant to the construction of optical instruments.

Opticians, Makers of Optical Instruments, and Experimental Natural Philosophers

Ordinary glassmakers, and indeed even some opticians, simply could no longer comprehend optical principles, which had become increasingly mathematicized during the late eighteenth century. In 1813, David Brewster reviewed the optimal procedure for opticians to measure refractive powers. First, the glass samples needed to be ground into a prism with at least two surfaces accurately plane and sufficiently polished. The angle of these surfaces then needed to be measured, and the refractive power of the prism determined from the observed deviation of the light ray. Brewster, however, admitted that this procedure was cumbersome and therefore rarely employed:

Even in determining the refractive power of different kinds of flint-glass for achromatic telescopes, the practical optician does not encounter the labour of forming them into prisms, but resorts to the easy, though inaccurate, method of estimating the index of refraction from the specific gravity of the glass.[38]

To obtain the refractive powers of a glass sample, Brewster proposed a more simplistic, yet just as accurate, method as the first one he described. For this procedure, neither grinding nor polishing the glass was necessary. The surface of the glass did not need to be regular, but could suffer from fractures. "It occurred to me," he noted, "that if a broken chip of any transparent solid were immersed in a fluid of the same refractive power, the incident rays would suffer no refraction in passing from the fluid into the solid, or from the solid into the fluid, and, consequently, that objects could be seen distinctly through the broken chip, whatever was the irregularity either of its form or of its surface."[39]

Brewster broke off a piece of crown glass with an irregular shape and dropped it into liquid Canadian balsam. The piece became nearly invisible in the liquid, and there was only a minute refraction of the light rays at the edges of the solid and liquid: the crown glass and the Canadian balsam had nearly the same refractive power. By mixing fluids of differing refractive powers, one could rather easily obtain a liquid compound that had the same refractive power as the solid. Now Brewster needed only a procedure to determine the precise instant when the refraction was totally obliterated at the edges of the two media, "as it is only in this particular state of things that the refractive power of the fluid could be regarded as a measure of the refractive power of the solid."[40]

Brewster placed a portion of the fluid whose refractive density was the closest to the solid between two lenses, *a* and *b*. (See fig. 6 in figure 2.2.) He then measured the distance, *bn*, at which an object at *n* was seen clearly. A small chip of the solid was then placed between the lenses along with the fluid. The rays diverging from *n* were then transmitted through the solid.

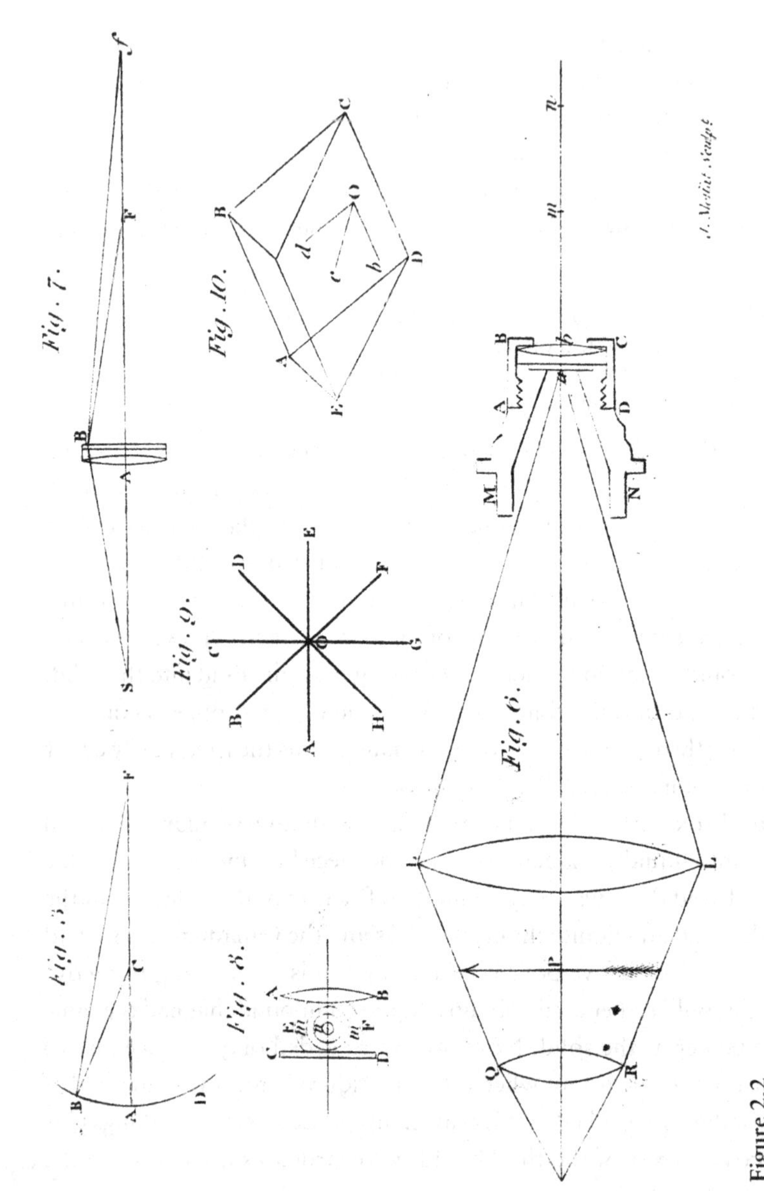

Figure 2.2
Sketch from David Brewster's *Treatise on New Philosophical Instruments* (1813) for determining the refractive powers of glass samples.

If the distance *bn* remains the same, then the solid and liquid have the exact same refractive power. But if the length *bn* is different, the refractive power of the fluid is altered until the object is seen clearly for the distance *bn*, both for the solid alone and for the solid and the liquid combined. The distance *bn* is now a measure of the refractive power of both substances. Brewster concluded that this method

> may be advantageously employed by the practical optician, for ascertaining the soundness and purity of the glass which he manufactures into lenses and prisms. There is, perhaps, no kind of labour more frequently wasted, than that which is employed in the formation of lenses and prisms of flint-glass. No sooner are the surfaces polished, than innumerable flaws and veins make their appearance, which the artist was before unable to discover, and which completely distort the image that is formed. A flint-glass prism, indeed, without veins and imperfections, is scarcely to be met with; and the amateur, who has tried to amuse himself in grinding the lenses for achromatic telescopes, must, for the same reason, have found it an impracticable attempt.[41]

Brewster then turned his attention to the measurement of refractive and dispersive powers of glass prisms, which was a vast improvement on Clairaut's and Boscovich's "variable top-angle prism" technique.[42] (The following is paraphrased.)

Let A be the angle of the fixed prism, whose refractive and dispersive powers are to be determined.

Let a be the angle of the variable prism, by which the refraction of the fixed prism is corrected.

Let α be the angle of the variable prism, which corrects the dispersion of the fixed prism.

Let R be the index of refraction of the fixed prism, while r is the index of refraction of the variable prism. dR and dr are the portion of the mean refraction to which the dispersion is equal.

Let $D = dR/(R - 1)$ be the dispersive power of the fixed prism.

Since $\sin(a - A)/R = \sin(a - x)$, which is the refractive power of the fixed prism, $\sin x$ can be determined easily. Therefore, $R = r \sin x/\sin A$. The dispersive power of the fixed prism is $\sin x' = (R/r)\sin A$.

Once x' has been determined, the dispersive power of the fixed prism and the portion of the mean refraction to which the dispersion is equal can be determined:

$$\frac{dR}{dr} = (R/r)\tan(\alpha - x')\cot(x'+1),$$

$$dR = dr(R/r)\tan(\alpha - x') + \cot(x'+1),$$

and

$$D = \frac{dr(R/r)\tan(\alpha - x')\cot(x'+1)}{R - 1}.$$

For these measurements, R and r are the indices of refraction for the mean refrangible rays for the fixed and the variable prism, respectively. dR is a part of the whole refraction and is always equal to the difference between the index of refraction for the first red and the last violet rays.[43] If R is approximately equal to r, then $x' = A$, and the equation for the dispersive power of the fixed prism can be simplified:

$$D = \frac{dr\tan(a - A)\cot(A + 1)}{R - 1}.$$ [44]

Such a technique of using the extreme rays of the spectrum was commonplace. It was not possible, until the work of Fraunhofer, to obtain a refractive index for each colored ray.

The method Brewster employed for obtaining the necessary variables of the above equation was rather ingenious (see fig. 2 in figure 2.3).[45] Prism *m*, whose dispersive power is to be determined, is placed on a ring. Prism *n*, the standard prism, whose refracting angles are such as to produce a greater dispersion than *m*, is fastened to a tube *fg*. As the circular head *AB* (which is divided into 360 degrees) is turned, so too turns the standard prism *n*, while the vernier on *ee'* remains fixed with prism *m*. A horizontal bar *AB* (fig. 4 in figure 2.3), approximately 3 to 4 inches wide, is stretched across a window such that it is perpendicular to a plumb line *CD* suspended from the top of the window. The instrument (fig. 2) is then placed far enough from the bar that a line joining the experimenter's eye at *O* and the center of the bar at *E* (fig. 4) is perpendicular to the anterior surface of prism *m* and to the bar *AB*. The instrument is turned until the section of the refracting surfaces of the prism *m* is perpendicular to *CD*, and is fixed in place by screw *s* (fig. 3). When *AB* is viewed though prism *m*, the bar's lower side will be bordered by yellow and red, the upper side by blue and violet. But after tube *fg* (fig. 2) with standard prism *n* is screwed onto the shoulder *ee'* (fig. 2), causing it to refract in opposition to *m*, the red and yellow fringes appear

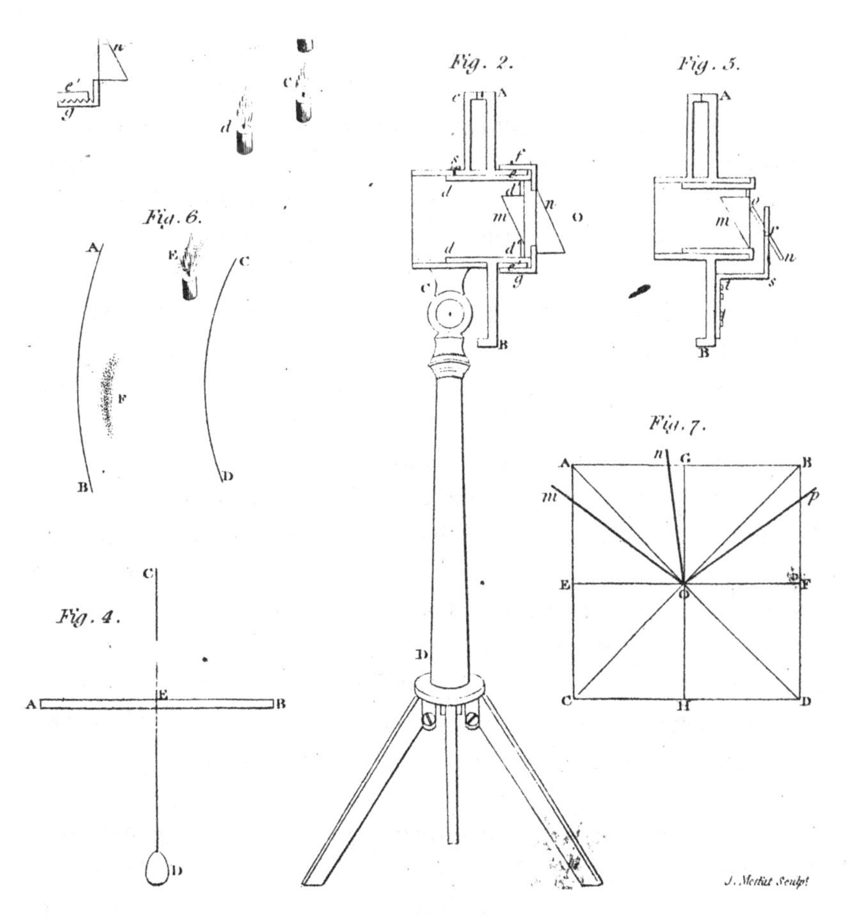

Figure 2.3
Sketch from David Brewster's *Treatise on New Philosophical Instruments* (1813) for determining the dispersive powers of glass prisms using the extreme rays.

above AB and the blue and violet fringes below. When AB is turned clockwise, the refracting angle of prism n diminishes, as do the color fringes. Once the colored fringes disappear, the degrees and minutes are read off the vernier. Turning AB counterclockwise until the colored fringes once again vanish, one notes as before the degrees and minutes when the bar appears free of color. If ϕ is the arch between those two angular positions, and B is the angle of the standard prism n, the refracting angle α that corrects the dispersion of prism m is $\alpha = \cos(\phi/2)B$. The dispersive power can now be obtained.

Using this technique, Brewster conducted a series of investigations on the correction of colored fringes formed by flint and crown glass.[46] From these experiments, he concluded that opticians should use flint glass with a dispersive power as low as possible and crown glass that has the least power of separating the extreme rays.[47] If two types of glass with similar dispersive powers were used, the flint glass would need lesser amounts of lead oxide, and therefore the glass would be less likely to suffer from striae and veins. In a footnote, Brewster asks: "As it is almost impossible to procure a piece of good flint glass more than 4 or 5 inches in diameter, might not a lens of any magnitude be composed of separate pieces of good glass from the same pot, firmly cemented together, and afterwards ground and polished?"[48]

Brewster's technique was more difficult for opticians to use than what William Hyde Wollaston had suggested some 11 years earlier, but it was much more accurate. Wollaston offered a convenient method for measuring the refractive indices, one that could be applied to the extreme rays of the spectrum.[49] His instrument (figure 2.4) consisted of a rectangular prism A of flint glass, below which the substance to be examined was mounted. BC was a rod or ruler 10 inches long, and CD and DE were each 15.83 inches in length. When the sights at B and C were placed such that the divisions between the light and dark portions of the lower surface of the prism could be seen through them, the rod F showed the index of refraction. The absolute refractive power of the substance was defined as[50]

$$\frac{(\text{Index of refraction})^2 - 1}{\text{Specific gravity}}.$$

This device determined the refractive powers by prismatic reflection of light at the inner surface of a dense refracting medium. The total reflection was

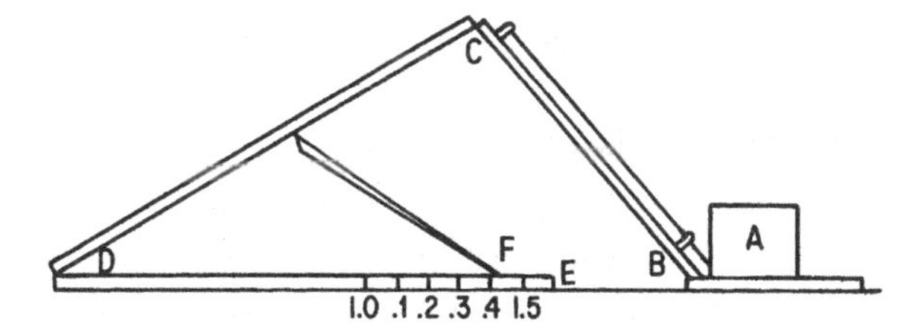

Figure 2.4
William Hyde Wollaston's instrument for measuring the refractive indices of substances. This instrument consisted of a rectangular prism of flint glass (A), under which was attached the substance to be examined. In this particular instance, the refractive index of the substance is 1.43.

determined by both the density of the reflecting prism and the density of the medium next to it. If the refractive power of one of the media was known, the other could be deduced by observing the angle at which a ray of light was reflected from it. The procedure was rather straightforward for the determination of the refractive power of fluids, but solids had to be juxtaposed to the flint-glass prism with a liquid or cemented with a higher refractive power than the medium being tested. Although the contraption was easy to use, Thomas Young concluded that Wollaston's measurements were accurate for the extreme red rays only, and Brewster claimed that the principle of prismatic reflection could not be applied with great precision to measurements of refractive power.[51]

In 1821 Herschel published his attempt to bridge the gap between the previous work by "celebrated geometers" (he named Clairaut, Euler, and d'Alembert) and optical instrument makers.[52] He complained that the result of the geometers' work was merely "a mass of complicated formulae" and that such formulae "have never yet been made the basis of construction for a single good instrument, and remain therefore totally inapplicable, or at least, unapplied, in practice."[53] Brewster echoed Herschel's sentiment by asserting that the works of "Euler, Clairaut and D'Alembert . . . have never yet been conducted to any practical or useful result."[54] Herschel even concluded that the work of Euler or d'Alembert actually retarded any developments of instruments by opticians, since the geometers claimed that it should be possible to determine the relative refractive powers of different media on rays of different colors. According to Herschel, this was experimentally impossible.[55] He also claimed that d'Alembert's idea to resolve spherical aberration for an object situated outside the optic axis was "carrying refinement too far. The difference of the [spherical] aberrations of an object-glass in and out of the centre of the field [the field is the plane of the telescopic axis and the object being viewed] is so small in ordinary telescopes, so to have escaped (so far as my enquiries have gone), the notice of the best practical opticians (and I have consulted many)."[56] Whereas the works of eighteenth-century savants should have led to the perfection of the telescope, they had in fact hindered its development:

> The investigations, from their dry and laborious nature, and the almost total want of that symmetry which is especially necessary in so complicated a subject, have been studied by few; the formulae, requiring a more extensive share of algebraical knowledge than can be expected in a practical optician, are thrown aside by him in despair, and the tables hitherto constructed from theory, being founded on data which may never again occur, are worse than useless, serving only to mislead. In consequence, the best and most successful artists are content to work their glasses by trial, or by empirical rules, embodying the result of numerous preceding trials, and which, therefore, have probably some analogy to that which would be the final results of theory, if presented in a tangible shape, and accommodated to the peculiarities of their constructions.[57]

Herschel then proceeded to render optical theory accessible to practical opticians by composing an essay on the geometrical optics relevant to lens making. He offered a basic derivation of formulae for the general theory of the aberration of spherical surfaces for rays incident in a plane containing the optic axis, and he detailed a straightforward method of correcting spherical aberration. He discussed chromatic aberration only briefly, arguing that refractive indices were a function of the "difference between [the length of] a ray of any assumed color and those most luminous rays in the spectrum"[58] and concluding that "little progress can be expected till more rigorous means have been devised of insulating the different homogeneous rays."[59] Fraunhofer provided such a "rigorous means."[60] It is therefore not surprising that Herschel was of the opinion that "however perfectly the foci of a double object-glass be adjusted to unite the extreme rays of the spectrum, a more or less considerable quantity of uncorrected colour remains, which cannot be destroyed by such adjustment."[61]

In 1822 Herschel published his second attempt to render geometrical optics accessible to opticians and instrument makers. It was an abstract of his 1821 *Transactions of the Royal Society* paper on the aberrations of compound lenses and objective lenses for "those who, without any large stock of mathematical knowledge, take a practical interest in the perfection of the telescope."[62] He continued by sympathizing with artisans:

> I am well aware how formidable a barrier is raised against improvements suggested by theory, by expressing them in a manner unintelligible to the many; and that, to the artist especially, the sight of an algebraic formula is apt to excite a degree of involuntary horror, a repugnance to come in contact with it, which no assurance of its correctness or utility on the part of its author is capable of overcoming.[63]

In consonance with his views as a mathematically oriented natural philosopher, Herschel wanted all artists to "construct a refracting telescope by regular rules, by any certain process, independent of trials."[64] He expressed his sentiment that artisans should build instruments and lenses from simple mathematical principles, rather than through the manual process of trial and error. He then proceeded to lay out the criteria for constructing achromatic lenses: The craftsman must unite the brightest red rays (bordering on the orange) with the brightest violet rays (next to the green). The extremities of the spectrum will deviate one way from the exact focus while the middle of the spectrum will deviate in the opposite direction, "thus producing the phenomenon always observed in well adjusted achromatic telescopes when thrown out of focus, viz. a purple or lilac fringe surrounding the image of a white object, on one side of the focus, and a green on the other [i.e., the secondary spectrum]. This is the criterion of a good adjustment of the foci; and to go beyond this point, with the ordinary materials, seems hopeless."[65] Hence, the artists were instructed to determine the ratio of the dispersive powers of flint and crown glasses, by working them into a small objective having the ratio of the focal lengths as close as possible in proportion to their dispersions. They should leave a preponderance on the side of the crown (convex) lens and reduce the curvature of one of its surfaces until the purple and green fringes surrounding a white object on a black background appear.[66]

Ironically, the standard for achromaticity was the generation of color, for Herschel rejected d'Alembert's attempts to destroy chromatic aberration for all colors—an attempt that, while being "refined in theory," was "useless in practice,"[67] as was d'Alembert's attempt to correct for off-axis rays.

As crude as Herschel's method might sound, it was nevertheless a vast improvement over earlier methods. Opticians in both Britain and France had previously ground and polished several glasses to the same radius of curvature, then compared their focal lengths when held up to sunlight.

Having done this for a specimen of crown and flint glass, the optician proceeded to compute his curves and "by his peculiar dexterity of tact, to work the faces to such perfect figures, that art can accomplish with given specimens of glass."[68] According to William Pearson, an influential member of the Astronomical Society of London well versed in optical theory and telescopes, this method was practiced by one of London's leading opticians, Charles Tulley, during the first two decades of the nineteenth century.

Herschel drew upon his own article in 1824 when he outlined the qualities desirable in flint glass for his report to the Joint Committee for the Improvement of Glass for Optical Purposes.[69] He drafted three criteria: achieve perfect transparency, improve homogeneity, and create a lens with a dispersive power between 1.667 and 2.000. Herschel continued by claiming that these criteria "permit the very ready and simple application of an exact mathematical theory to the construction of object glasses without supposing any knowledge of mathematics in the artist beyond the working [of] the rule of three sums."[70] Herschel was concerned with developing mathematical principles that a worker could apply. According to Herschel, rational mechanics and arithmetic communicated in written, algorithmic form would result in the production of superior lenses.

In 1822 Herschel applied his method of extreme rays to crystal apophyllite, measuring the angles of refraction for the extreme red and violet.[71] However, in a letter to Brewster which was subsequently published, Herschel detailed his interest in the full range of colors in the prismatic spectrum[72]—as had nearly everyone in France, since it was a pressing theoretical issue. Since 1819 Herschel had been interested in tints produced by crystals in polarized light. He measured the diameter of the rings of different homogeneous colors, noting that he had

> extreme difficulty . . . procuring a tolerably homogeneous ray by this means; owing to the sun's diameter, the irradiation of the sky, *and the imperfections of prisms,* not to speak of the mobility of the spectrum, being then unprovided with a heliostat, I began to study the resources afforded by transparent coloured media, in hopes that I might discover some, whose limits of transmission, either simply or combined, might be such as to allow the passage of rays only within very narrow limits of refrangibility. In this examination I immediately encountered the singular phenomenon, noticed, I believe, first by Dr. Young, of an almost total obliteration of some of the colours by certain glasses, while others intermediate between, or sharply bordering on those obliterated, appeared to be transmitted in all their brilliancy, thus producing an image (when a narrow luminous line is examined with a prism, and such a coloured glass), not consisting of a broad band of gradually varying colour,

but an assemblage of more or less sharply defined coloured streaks of different breadths and colours, separated by intervals, in some cases, absolutely black, in others only feebly illuminated.[73]

With a piece of blue glass containing a tinge of purple—commonly manufactured for sugar basins and finger glasses—Herschel analyzed the color separation. The red portion of the spectrum was divided into two sections. The less refracted portion was well-defined, perfectly homogeneous light, and was separated from the other portion by a wide band of black. This other red was not homogeneous. "Its place in the spectrum is such, that its most refracted limit exactly comes up to that remarkable black line which Dr Wollaston observed separating the red from the yellow in the solar spectrum."[74] Fraunhofer's work was not mentioned. Herschel's paper was read before the Royal Society of Edinburgh in November of 1822 and therefore itself marks a sort of boundary: this was his last paper on the colors of the spectrum and their isolation before he visited Fraunhofer.

Herschel provided an experimental technology for using colored flames and media to measure the dispersion of the extreme rays.[75] He argued that the advantage of using the properties of colored flames and media for optical research was that one could isolate rays of several species with a very definite refrangibility, which could be used for a direct determination of the dispersive power of the media on the extreme rays and a more accurate calculation of the dispersive power of media on the intermediate ones.[76] The method he employed was as follows. (See figure 2.5.) He cut two parallel slits in a screen (*AA*, *BB*). *AA* was twice the length of *BB*, and each was 0.2 inch wide. This screen was then placed horizontally in a window. A prism of a known refracting angle was used to examine the light entering from the slits. The eye of the investigator peered through a blue lens whose thickness was sufficient to stop the green, yellow, and most of the refrangible red rays. Two images of each slit were seen: a red and a violet one for each (*aa*, *a'a'*, *bb*, *bb'*). As the prism was pulled away from the investigator's eye, the violet image *a'a'* of the longer slit approached the red *bb* of the shorter, and indeed appeared to overlap it. At one point (the prism being at a minimum deviation) the corresponding edges *a'a'* and *bb* formed a single, straight line. The distance between the prism and the slit was then measured. The difference of deviation of the extreme rays was equal to the angle subtended by the distance between the corresponding edges *AB* of the slits. Once this was known, the dispersive power could be determined:

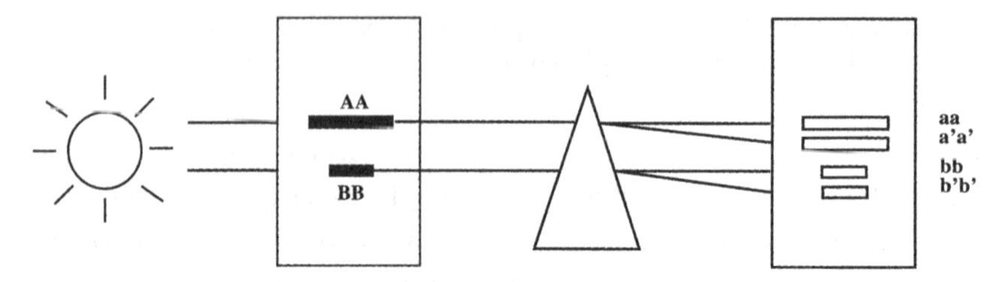

Figure 2.5
John Herschel's method for determining dispersive powers of a glass prism using the extreme rays.

i = distance between corresponding edges of slits observed;

d = distance from prism to slits;

A = angle of prism;

D = deviation of the extreme red rays;

μ = refractive index for those rays;

$$p = \text{dispersive power of lens} = \frac{\partial\mu}{\mu - 1},$$

where

$$\partial\mu = \frac{i}{2d}\cos\left(\frac{A+D}{2}\right)\sin\left(\frac{D}{2}\right).$$ [77]

Herschel concluded his article by describing how the dispersions were obtained experimentally:

It will be recollected, that they are founded on observations of rays situated rigorously at the extremities of the spectrum. These rays elude all ordinary observation in the solar spectrum, and are too feeble to exert any sensible influence on the colours of the edges of the objects, in the usual mode of compensation. This latter, indeed, being merely comparative, assuming as known the dispersion of a standard prism, its results must be affected by all the uncertainties attending the determination of this element, which, if obtained by actual measurement of the solar spectrum, must, as I have before observed, necessarily err considerably in defect: add to this that the method of compensation, owing to the "irrationality of the coloured spaces,"[78] can only give results corresponding to the union of the two brightest and most strongly contrasted colours, which may differ considerably from those corresponding to the extreme rays.[79]

In a paper published in 1822 in the *Edinburgh Philosophical Journal*, Herschel discussed the production of achromatic lenses for telescopes and the calculation of refractive and dispersive indices.[80] He informed his reader that the refractive powers of glass could be determined by grinding a small portion of the glass into a prism or lens and then observing the deviation of the most luminous rays. For the determination of the dispersive power of a refracting telescope, "the best we can do is to work the lenses [of flint and crown glass] so as to produce the same compound focus, *not for all rays, for that is impossible*, but for the two brightest and strongest colours in decided contrast with each other."[81] Herschel once again recommended, this time as a result of his work on crystals, that the two colors to be united should be the brighter red, bordering on orange, and blue, precisely where it begins to pass into green. He suggested that the optician

> determine the ratio of the dispersive powers of his flint and crown glasses, by a direct experiment on small portions of his materials, working them into a small object-glass, having the ratio of the focal lengths of its component lenses, as nearly as he can guess, in the proportion of their dispersions, but leaving rather a preponderance on the side of crown or convex lens, and then by degrees reducing the curvature of one of the surfaces of this, till he obtains the nearest possible approach to perfect achromaticity, *i.e.*, till the purple and green fringes surrounding a white object on a black ground [the secondary spectrum], appear in it as above described, when thrown one way or the other out of focus. . . . Let him then determine accurately, *by experiment*, the focal length of each of his two lenses, and dividing the one by the other, he will obtain a *dispersive ratio* (ratio of the dispersive powers), on which he may calculate *with perfect security* in his future operations. If he know [*sic*] the exact radii of his tools, he may at the same time determine the refracting powers of the media.[82]

Peter Barlow continued Herschel's attempt to apply well-developed optical theories to the production of achromatic telescopes for the Board of Longitude.[83] He began his article by echoing Herschel's astonishment over the lack of communication between theoreticians and opticians: "It is very remarkable, since the achromatic telescope is altogether of English origin, that in no one of our separate optical treatises are to be found specific rules for its construction, fitted for the use of practical opticians."[84] He added that there were, however, numerous attempts made by foreign mathematicians.[85] This paper, published in 1827, reflected the post-Fraunhoferian point of view. One no longer finds the claim that refractive and dispersive indices cannot be determined for each color. In fact, in his essay of January 1828 on fluid-lens telescopes, Barlow used Fraunhofer's refractive indices for crown and flint glass for the red, green, and violet rays.[86] George Biddell Airy, at the time the Lucasian professor of mathematics at Cambridge, incorporated Fraunhofer's values for the refractive indices of three types of glass (two flint and one crown) for the *B*, *E*, and *G* Fraunhofer lines in his essay on the construction of achromatic eyepieces of telescopes.[87]

Fraunhofer and Textbook Optics

One of the most widely used university textbooks on optics in Britain during the 1820s was the Reverend Henry Coddington's *Elementary Treatise on Optics*, first published in 1823.[88] This treatise was based on some physics lectures by William Whewell that Coddington had attended as a student at Trinity College, Cambridge, in the spring of 1819.[89] In the section on chromatic aberration, Coddington discusses the basic trigonometric equations used in explaining the phenomenon: "We suppose here that there are seven distinct parcels of colours, each refracted to its proper focus, but as in the most perfect experiment of the prismatic spectrum, there are no intervals between the colours, the number of foci should probably be infinite."[90] Coddington continues by explaining how chromatic aberration can be quantified for a single lens (figure 2.6): *Bv*, *bv*, *Br*, and *br* are the extreme violet and red rays from opposite points of the lens. The line *nmo* joins the intersections of *Br* and *bv* and of *Bv* and *br*, respectively. Thus, the line segment *no* is the diameter and *m* the center of the circle of aberration. (The circles for all other colors in respect to red lie on the cone *nmr* and so are smaller than *nmo*.) Simple geometry yields

$$no = Bb\left(\frac{Ar - Av}{Ar + Av}\right).$$ [91]

Let $1 + r$ and $1 + v$ be the indexes of refraction belonging to the red and violet rays, respectively. Given that

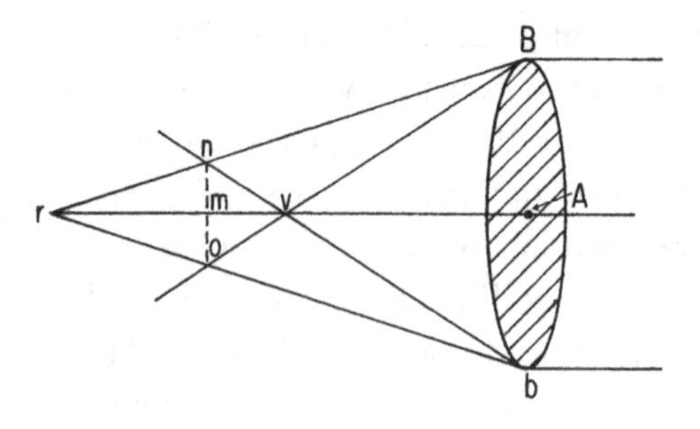

Figure 2.6
Henry Coddington's quantification of chromatic aberration for a single lens, 1823.

$$(1/F) = (m - 1)/p,$$

where F is the focal length, m is the index of refraction, and p is the radius of the lens, we then have

$$Ar = p/r,$$

$$Av = p/v,$$

and

$$\frac{Ar - Av}{Ar + Av} = \frac{v - r}{v + r}.$$

Let β be the aperture *Bb*. Then

$$no = \frac{\beta(v - r)}{v + r}.$$

This last equation can now be used as a guide for calculating a refractive index given the circle of aberration and the other index, or for calculating a circle itself, given two indexes. For example, if a lens of crown glass has $v = 0.56$ and $r = 0.54$, the diameter of the greatest circle of aberration is 1/55 the aperture (*Bb*).[92]

Coddington then discussed the problem of constructing an achromatic lens and the attempts to remedy the problem by joining together lenses of equal and opposite dispersive indices. The way to resolve the problem of achromaticity was to correct the extreme red and violet rays of the spectrum. If $1 + r$ and $1 + v$ are the values of m (or indices of refraction for the red and violet rays respectively), then, taking two lenses, where ϕ is the focal length of the combined lenses and the primed values are for the second lens, one gets

$$(1/\phi) = (r/p) + (r'/p')$$

for the red rays and

$$(1/\phi) = (v/p) + (v'/p')$$

for the violet rays. Equating the two values yields

$$(r/p) + (r'/p') = (v/p) + (v'/p');$$

$$rp' + r'p = vp' + v'p,$$

or

$$\frac{p'}{p} = -\frac{v' - r'}{v - r}.$$ [93]

Thus, in order to unite the most unequally refrangible rays (i.e., the extreme red and violet rays) to one focus, a convex and a concave lens must be combined to render the ratio of each lens proportional to its corresponding dispersive power. Flint and crown glass, which were commonly used in this way, had dispersive powers in the ratio of 50 to 33.[94] However, flint and crown lenses do not disperse the different colored rays proportionally. Therefore, if the extreme rays of red and violet are corrected, the green and blue rays will remain uncorrected and chromatic aberration will result.

Coddington's treatise was the last optical text to appear in Britain before the British learned about Fraunhofer. Fortunately, Coddington published a second edition in 1829.[95] It was the first British textbook on optics to be published after Fraunhofer became well known among British experimental natural philosophers. Coddington wrote in his preface to the new edition:

> The Science of Optics having of late years assumed almost a new form, it has appeared desirable that a Treatise should be drawn up, by which Students might be led, with the least possible difficulty, to the comprehension of those important Theories which have extended all, and superseded many points of the subject as contained in former works written for their use.[96]

One of the reasons for the transformation in optics during the mid 1820s was Fraunhofer's research. Coddington wrote of Fraunhofer:

> These [spectral] interruptions, first observed imperfectly by Dr. Wollaston, and afterward independently, and with great precision, by Professor Fraunhofer of Munich, and by him termed the *fixed lines* in the spectrum, are one of the most important discoveries in the whole range of Optical science.[97]

Coddington incorporated Fraunhofer's alphabetical designation of the spectral lines as well as his calculations of refractive indices of differing substances at those particular fixed lines.[98] He added that although the world was indebted to Brewster's work on the total dispersion of a glass prism, or the difference between the greatest and least deviation of a prism, $\Delta(\mu - 1)''$, the experimenter should, unlike Brewster, "observe, as Fraunhofer did, the refraction of each fixed line."[99]

Fraunhofer's method was employed in Coddington's text exercises concerning the chromatic aberration of two lenses placed in contact. Following Herschel's suggestion, Coddington writes, the optician should unite the *D* and *F* Fraunhofer lines, since "the exact union of these will ensure the approximate union of all the rest, better, on the whole, than if we aimed at uniting the extremes of the spectrum, and a far greater concentration of light will be produced."[100] In an attempt to combine two lenses to produce an achromatic union, two variables, the focal lengths of both lenses, needed to be determined by means of two equations: one that assigned the focal length of the combination of lenses, the other that provided the focal length of two united rays demarcated by the Fraunhofer lines. A third lens would unite the two unified rays with a third ray. In short, Coddington incorporated the earlier methods provided by both Brewster and Herschel[101] for the determination of dispersive and refractive indices, but with Fraunhofer's critical modification.

Brewster also included Fraunhofer's contribution to optics in his *Treatise on Optics*, published in 1831:

> One of the most important practical results of the discovery of these fixed lines in the solar spectrum is, that they enable us to take the most accurate measures of the refractive and dispersive powers of bodies, and by measuring the distances of the lines *B, C, D,* etc. Fraunhofer computed the table of the indices of refraction of different substances. . . . From the numbers in the table here referred to we may compute the ratios of the dispersive powers of any two of the substances, by the method already explained in a preceding chapter.[102]

Brewster included Fraunhofer's table of the refractive indices of several glasses and fluids for the Fraunhofer lines *B* through *H* to the nearest sixth decimal place. And he provided the indices of a crown-glass prism for each of the colored rays to the nearest fourth decimal place.[103] As in Coddington's second edition, and not in Brewster's earlier work, *A Treatise on New Philosophical Instruments* (1813), the recommended procedure for determining the refractive index of a sample is no longer taking the arithmetic average of the extreme rays, but determining the indices using the dark lines to demarcate precise portions of the solar spectrum. Fraunhofer had left an indelible mark on geometrical optics.

3

Artisanal Knowledge and Achromatic Lenses

These things [skills affiliated with building musical instruments] cannot be learned from books. One rather needs to see them and be shown them by a good master so that he can teach the necessary manual techniques.

—Georg Andreas Sorge, 1773

Fraunhofer engaged in his secretive research on optical glass behind the protective walls of Benediktbeuern, a secularized Benedictine monastery nestled in the foothills of the Bavarian Alps. Guild and monastic secrets that ensured the financial supremacy of the Optical Institute lay at the heart of his optical expertise. For nearly 100 years after his death, only a privileged few were permitted access to the highly restricted and private techniques of optical glass manufacture.

With this chapter I seek to consider Fraunhofer's enterprise in its full scientific, economic, political, and social context. I wish to localize the Fraunhofer story within the framework of the Napoleonic Wars and the subsequent French occupation of Bavaria by discussing the interaction of differing cultures during the early nineteenth century—an interaction that enabled Bavaria (and, later in that century, Germany) to proclaim cultural and scientific hegemony.

In the first section of the chapter I shall discuss the context of Fraunhofer's manufacture of optical glass. In 1801 Napoleon ordered the Bureau topographique to provide accurate topographical maps of his new ally, Bavaria. New, more accurate maps enabled King Maximilian I to base his new taxation scheme on the amount of property his subjects owned. This tax was used to bolster his plans for reform, as previous maps had been too crude and inaccurate for such purposes. The Optical Institute, where the young Fraunhofer was an apprentice, originally supplied achromatic lenses for ordnance surveying (Landesvermessung) and astronomical instruments for the Bureau topographique and the newly created Land Registry, which was responsible for determining property boundaries. It was the site of Fraunhofer's now-famous Six Lamps Experiment and his Solar Light Experiment as well as the site of his production of the world's most coveted achromatic lenses. The secularization of Bavarian monasteries in 1803 provided Fraunhofer with both a huge space, where he conducted large-scale experiments necessary for the detection of spectral lines, and a skilled labor force, in the form of Benedictine monks and artisans from local guilds, well versed in optical theory, physical instruments, and glass manufacture. As Fraunhofer's reputation spread throughout Europe, the Optical Institute began to manufacture achromatic lenses for the instruments of leading astronomers. It was housed in Benediktbeuern from 1807 to 1819.[1]

The second section of the chapter analyzes the content of Fraunhofer's optical work. By understanding his particular, local context, one can more fully appreciate his artisanal expertise. He required the large space that a monastery could provide in order to conduct his Six Lamps Experiment. He also needed a space that had various degrees of privacy already physically built in. Although the Optical Institute was originally established as a state institution, Joseph von Utzschneider transformed it into a profit-seeking business. Hence, strict measures were implemented to ensure that no information concerning the manufacture of achromatic glass would leave the confines of the monastery. Fraunhofer rendered certain portions of his enterprise public, such as the publication of the method using the dark lines as a calibration technique for producing achromatic lenses, while keeping other portions private, such as the method and recipe for manufacturing those achromatic lenses. Fraunhofer's choice of what to disclose and what to keep secret was very clever. By convincing prospective clients (both entrepreneurial and scientific) that his method of calibration (based on the spectral dark lines) was the most accurate in existence for producing lenses, and by demonstrating with that particular method that his lenses were indeed of superior quality, Fraunhofer immediately increased the market for his institute's products. By neither publishing the technique for glassmaking nor permitting access by experimental natural philosophers to the Glashütte where he made his optical glass, Fraunhofer ensured that the ever-increasing market would be compelled to purchase only his institute's lenses; he guaranteed the market's fidelity.

Fraunhofer's procedures for procuring achromatic glass and his ability to utilize the Benedictine architecture (itself a manifestation of the Rule of St. Benedict's emphasis on labor, silence, and secrecy) in order to execute his experiments form another part of the discussion. Secrecy was important to the Benedictine monks and guild artisans, who were Fraunhofer's assistants. The monks possessed a thousand-year tradition in the manufacturing, cutting, and polishing of glass, particularly stained glass for cathedrals.[2] Monastic texts claimed that glass was the manifestation of the divine light and the light of truth (lux divinus and lux veritatis) on Earth. These texts even drew an analogy between glass and the Holy Spirit.[3] Knowledge of glassmaking was said to lead to an understanding of God and his relationship with humankind. In addition to this epistemological reason for glassmaking, there was an economic factor: monasteries prospered by making stained glass for town halls and for the private residences of aristocrats and wealthy merchants. Until their secularization, monasteries had owned a large number of glasshouses throughout Bavaria, Bohemia, and the Swiss cantons.[4]

Fraunhofer's work on the dark lines of the spectrum, as we shall see, was not the culmination of any concern on his part with the nature of light. Nor did he use those lines to analyze substances. Rather, his work was a product of artisanal training with a view to perfect the construction of achromatic lenses for astronomical instruments such as telescopes and heliometers, as well as ordnance surveying instruments such as theodolites. Looking back 10 years to the period when he had worked on calibrating achromatic lenses, Fraunhofer remarked:

> There were not until the present time any fixed theoretic principles for the construction of achromatic object glasses: and opticians were obliged, to a certain degree, to rely on chance, which made them polish a great number of glasses, and select those in which the faults were most compensated. As the probability of this chance is much less in large glasses than in small ones, even those of the middle size would have been seldom perfect; and even with the best flint glass, the construction of large achromatic object glasses would have been impracticable. The more important causes that rendered this process necessary are as follows: that the theory of achromatic object glasses is as yet imperfect; that the means formerly applied for ascertaining the powers of refraction and dispersion of colors in the different species of glass, which ought to rest on a firm basis, are not sufficiently established; and because the methods hitherto used for grinding and polishing the glasses are not being calculated to follow the theory with the degree of exactness that they ought, if a palpable indistinctness should be avoided.[5]

Fraunhofer's early work, including his influential experiment involving six lamps and solar light, was not an attempt to vindicate or debunk either the wave or the corpuscular theory of light. More than anything else, Fraunhofer's work was based on, and bound to, his technical mastery of artisanal knowledge and skill. Fraunhofer's knowledge of geometrical optics, reflected in both the experimental design and the calculation of the refractive indices for his glass prism samples, was in fact quite basic. Indeed, his purposes required only knowledge of (and practical skill in handling) Snel's Law of Refraction, and basic trigonometric relations.[6] Fraunhofer deployed Snel's Law and its implications in his everyday work with glass prisms.

Before Fraunhofer's work, the determination of dispersive indices of glass prisms was restricted to the rather imprecise procedure discussed in the previous chapter, whereby the optician would take the mean of the refraction of either the extreme red and violet rays or the two brightest and most strongly contrasting colored rays. This two-ray method had been necessary because prisms had not generally been good enough to demarcate precise portions of the solar spectrum. In such prisms, red would blend into orange, orange into yellow, and so forth. Fraunhofer, as we shall see, was able to determine the refractive indices for a much more precise portion of the solar spectrum by using the solar dark lines as boundary markers. Only his prisms—produced by means of skilled manipulations unparalleled elsewhere—could show such a large number of lines. Extremely pure ingredients, special stirring techniques to ensure homogeneity, and a novel method for polishing lenses perfectly symmetrically yielded Fraunhofer's first line-displaying prisms, which were subsequently used as calibration devices in producing even more accurate prisms and lenses. This story, in part, deals with the invention of a method for testing quality that, once widely accepted, gave Fraunhofer and his Optical Institute a significant lead time for perfecting glassmaking techniques.

The Context of Fraunhofer's Optical Work

Napoleon, the Bureau topographique, and the Ordnance Surveying of Bavaria

After sustaining losses in western and southern Germany during the last two years of the eighteenth century, French armies were revitalized by Napoleon's return from his campaign in Egypt and his self-proclamation as first consul of the Republic. In 1800 Napoleon's forces orchestrated a series of victories over the poorly organized Austrian army. The culmination of those victories was the devastation of the Austrians in the Hohenlinden forest, several miles east of Munich, on 3 December 1800.[7] The ensuing Treaty of Lunéville, signed by France and Austria in February of 1801, strongly affected Bavarian history.

Bavaria's government, led by Prince Elector Karl Theodor until his death in February of 1799, had played a subservient role to Bavaria's more powerful southern neighbor, Austria. Indeed, many Bavarians had blamed Karl Theodor's mismanagement of the government on his Austrophilic tendencies.[8] The ascendancy of Prince Elector Maximilian Joseph IV to the throne not only signaled a change in Bavarian foreign policy, but also brought about aggressive economic and social reforms and major scientific development. In February of 1799 Maximilian Joseph IV proclaimed: "I was raised in France and request that you consider me a Frenchman. The joy that I found when I heard a follower of the arms of the Republic proved to me that I am a Frenchman."[9] Maximilian Joseph IV and his administrative reformer, Maximilian Joseph Freiherr von Montgelas, were educated in Enlightenment France and remained strongly francophilic to their deaths.[10] They intended to emulate the French in order to replicate France's greatness.

During the negotiation of the Treaty of Lunéville, the French foreign ministry quite cleverly leaked to the Bavarian press that Austria still hoped to annex Bavaria. This was shocking news to the Bavarians. After all, they knew very well that France was the only European power that supported Bavaria's claims for land compensation from Prussia. The French agreed to compensate Bavaria for the property to the west of the Rhine which the Wittelbachers, the ruling family of Bavaria, had lost. France even renounced Bavaria's huge debt. France's shrewd diplomatic stance, coupled with Bavaria's new francophilic government, resulted in a Franco-Bavarian agreement in August of 1801, which was solidified by an official alliance in the fall of 1805.[11] The resulting French impact on Bavaria was immense, for it expedited Bavaria's economic evolution by establishing a network between Paris and Munich that enabled political and scientific topics to be discussed and vast transfers of technology to take place.[12]

One technology transfer concerned the Bureau topographique français. In 1800 Napoleon established his Commission des Routes, which was directed by Charles Frérot d'Abancourt and which included the former Bavarian privy councilor and Geheim-Finanz Référendaire, Joseph von Utzschneider. Napoleon ordered this commission to create a topographic map of Bavaria, and the prince quickly advised the Bavarian administration to comply with Napoleon's request. On 19 June 1801, Maximilian Joseph IV and Montgelas set the project in motion.[13] Twenty-four Bavarian and nine French geodetic engineers and surveyors worked in unison to produce a "topographically and astronomically correct map."[14] Ordnance surveying projects, which were being undertaken throughout central Europe during the late eighteenth century as a result of the Napoleonic Wars, required detailed knowledge of optics. Land distances were determined by series of triangulations. A base line for each triangulation project was created by using wooden rods of a known length placed one in front of another and stretching for kilometers. Other land distances could be measured much more easily by determining the angle subtended by the base line with another point kilometers away. Theodolites and repeating circles were capable of measuring this angle rather precisely.

Since Benedictine monks had the most comprehensive optical knowledge in Bavaria during the eighteenth century, they were the individuals most often hired to undertake ordnance surveying projects throughout that region. Monastic astronomers were also needed, as the instruments and the geometrical calculations used in ordnance surveying were similar to (and in some cases the same as) those used in astronomy. The Bavarian government drew upon these monastic and bureaucratic resources in order to create a topographical map for Napoleon and his army of occupation. The surveying instruments used by the Bureau topographique were manufactured by Fraunhofer's Optical Institute and tested on site by those individuals responsible for the ordnance surveying of Bavaria.

Utzschneider played a critical role in the Franco-Bavarian ordnance surveying project. On 22 August 1800, after Napoleon's order to the Commission des Routes for a topographic map of Bavaria, Maximilian Joseph IV appointed Geheim-Finanz Référendaire Utzschneider the financial liaison between the Bavarian government and the French général adjutant commandant C. F. d'Abancourt.[15] Utzschneider was ordered to pay d'Abancourt 9000 francs for the initial cost of starting up the Bureau topographique in Bavaria.[16] He was also responsible for hiring opticians and astronomers from the area.

Utzschneider turned to the French for guidance. The enterprise was headed by the engineer-geographer and directeur des opérations géoditiques, Colonel Bonne. The French were the world's most experienced ordnance surveyors. In 1635, King Louis XIII and the founder of the Académie royale des sciences, Cardinal Richelieu, had both recognized the importance of geodesy. Geodesy not only was necessary for astronomical calculations, it was also needed to assist the disciplines of topography and cartography. As a result of such obvious utility to the state, the Académie had funded the famous triangulation project between Paris and Dunkirk by Abbé Jean Picard, Jean-Dominique Cassini (I), and Philippe de La Hire, completed in 1690. From that time on, geodesists performing ordnance surveying enjoyed a lucrative relationship with the Académie and the French monarchy. Ordnance surveying was part of the curriculum of the French Military Academy at Metz where military engineers and geographers, such as Bonne, were educated and trained. Bonne was assisted by two Bavarian officers, Colonel Adrian von Riedl and General Müller.[17]

Utzschneider appointed the Benedictine monk Ulrich Schiegg—also a professor of mathematics, physics, and astronomy at Salzburg—Hauptastronom (chief astronomer) of the ordnance surveying section of the Bureau topographique and astronomer of the Munich Observatory after his monastery was secularized in 1803.[18] Schiegg had had previous experience in ordnance surveying; from 1784 to 1790 he had led a surveying team that mapped out the property of the Ottobeuern monastery.[19] Joseph Niggl, the Bavarian optician who at one point worked with Schiegg and was the optician to the Rott monastery's observatory, was appointed to the project in 1804.[20] He was responsible for the cutting of glass lenses for the ordnance surveying instruments, for which he received a sum of 50 florins.[21] Niggl, like many young optical apprentices from Bavaria, learned his trade in the Benedictine monastery at Rott, where he manufactured reading lenses for the monks.[22]

On 26 February 1808, at the request of Schiegg and Utzschneider, Johann Georg von Soldner accepted a position in the Bavarian ordnance surveying project.[23] Soldner had been a Prussian officer and director of the ordnance surveying of Ansbach. After the Prussian defeat by the French at the battle of Jena and Auerstedt on 14 October 1806, Soldner found himself unemployed. Schiegg encouraged Soldner to consider coming to Bavaria, where a much larger ordnance surveying community already existed; Soldner quickly accepted the position. His most famous contribution to the Bavarian project was his improvement of techniques for the calibration of areas under spherical triangles in his 1810 treatise *Theorie der Landesvermessung*.[24] Jean-Baptiste Delambre's previous calculations had been based on linear triangles. But since distances of triangulation are quite large, the error introduced by the spheroid-shaped Earth becomes significant.[25] Soldner's method of calculation, which was subsequently improved upon by Carl Friedrich Gauss when he employed a much more advanced form of mathematics during his ordnance surveying of Lower Saxony, immensely enhanced the accuracy of ordnance surveying.[26] Schiegg instructed his students in Soldner's method of triangulation.[27] From 1815 to 1818 Soldner served as an advisor to Georg von Reichenbach of the Mathematisch-mechanische Institut and Utzschneider by testing the precision of the Optical Institute's theodolites, whose lenses were produced by Fraunhofer.[28] The French stayed to assist the Bavarians until 10 March 1808, when they departed to occupy newly acquired Prussian territories.[29] King Maximilian I reformulated the Bureau topographique, now called the Topographisches Bureau.[30] It was still based on the same organizational structure that the French had originally established in which astronomers, engineers, and geographers coordinated efforts for the triangulation project. The king, however, reduced the total number of employees in the enterprise.[31]

In 1801, the year of the establishment of the Bavarian ordnance surveying project, Utzschneider created a Bureau de cadastre (Bureau of Land Registry) within the Bureau topographique in Munich. He reckoned that one could more accurately determine rates of property taxation using the new measurements of the region determined by the ordnance survey.[32] It had been decided on 5 October 1799 that taxation was the most effective way for Bavaria to balance its budget and to reform its government. Utzschneider employed many individuals associated with the ordnance surveying project, including Schiegg and Soldner, to provide accurate mappings of properties to prorate the tax schemes.[33] These maps had to be much more precise than the general topographical map requested by Napoleon for his troops. Before Utzschneider's efforts, there were 114 different basic taxes (Grundsteuern), regulated by a variety of grossly antiquated taxation schemes.[34] The joint Bureau de cadastre and Steuervermessungskommission (Commission for Tax Determination) gave rise to the Unmittelbare

Königliche Steuerkatasterkommission (Royal Tax Commission) by order of King Maximilian I on 27 January 1808.[35] On 12 April of that same year, Schiegg received a royal appointment as "Instructor of Geometers and Geodesists for the Measurement of Tax in Royal Bavaria."[36]

The Bavarian joint ordnance surveying and tax determination enterprises were amazingly successful. In 1811 Reichenbach traveled to Paris to show the French geometers and astronomers his instruments used by the Bavarian ordnance surveying project. In front of the French Bureau des longitudes, Reichenbach impressed his French audience (which included Pierre Simon Laplace, Joseph-Louis Lagrange, Jean Baptiste Biot, and Dominique Francoise Arago) with a meridian circle, an equatoreal, and a flint glass object lens 6 Parisian inches in diameter (about 6½ English inches, or 162 millimeters) made by Fraunhofer. The members of the Bureau des Longitudes were all convinced that the Bavarian ordnance surveying and optical technology surpassed even the most stringent French criteria of optical excellence.[37]

The Mathematisch-mechanische Institut

Both the ordnance surveying enterprise and the Bureau of Land Registry required a large supply of surveying instruments and an instrument repair shop. Not coincidentally, in 1801—the same year the two projects were created—Reichenbach and Joseph Liebherr began the groundwork for the creation of the Mathematisch-mechanische Institut. The institute, jointly funded by the Royal Academy of Sciences in Munich and the Bureau topographique, provided the Bureau topographique and the Bureau de cadastre with theodolites, telescopes, and repeating circles.[38]

Reichenbach had shown an aptitude for machines at a very early age. On 16 June 1791, at the age of 19, he had traveled to England in order to obtain firsthand knowledge of instrument building. He had spent his first few months in Matthew Boulton and James Watt's company, noting the details of their devices—particularly Watt's steam engine. As it happened, Reichenbach played a crucial role in the Bavarian usurpation of Britain's optical instrument industry. He had met with the British instrument maker Jesse Ramsden twice in order to discuss astronomical instruments, particularly theodolites. Upon his return to Munich in 1796 he had found "not a single establishment for the manufacture of mathematical, let alone astronomical instruments. All such instruments were supplied from England, or under extreme circumstances from the workshops of Hrn. [Georg Friedrich] Brander or Hrn. [Caspar] Höschel from Augsburg."[39]

In 1801 Reichenbach joined forces with a Munich clock maker named Joseph Liebherr to manufacture a device used in dividing a circle into degrees, minutes, and seconds.[40] A year later, the French geographical engineer and former director of the Mannheim Observatory, Henry, who also worked on the Franco-Bavarian ordnance surveying project, convinced Reichenbach and Liebherr to establish a full-fledged institute devoted to mathematical and astronomical instruments which could supply the Bureau topographique.[41] Schiegg served as advisor to Reichenbach on the construction of geodetic and astronomical instruments, particularly an 18-inch astronomical circle and a 16-inch azimuth circle.[42] Schiegg would offer his practical suggestions after testing the equipment on site for the triangulation procedures.

In 1804 Utzschneider convinced Reichenbach and Liebherr that the institute needed a financial advisor. On 20 August, Reichenbach, Liebherr, and Utzschneider signed a contract making them co-directors of the Mathematisch-mechanische Institut.[43] Utzschneider recalled some 20 years later:

> [Reichenbach and Liebherr] expressed their desires for me to expand the workshop's scope and to create an ordinary institute for large and small instruments and machines, such as those produced in England. I did not refuse to enter into an agreement with them not only for this reason, but also because such an institute could produce young, skilled mechanics, which Bavaria lacks.[44]

This joint venture was divided into three sections. Reichenbach directed all technical aspects of the institute and manufactured the instruments. Liebherr was the Meister of the institute; he executed Reichenbach's plans and taught the young journeymen. Utzschneider produced the required funds and oversaw the institute's entrepreneurial interests.[45]

The shortage of skilled labor certainly affected Utzschneider's search for lensmakers. The blockade separating Great Britain from the Continent in 1806 also augmented the demand those craftsmen. It resulted in the disappearance of British lenses from European markets (optical lenses had previously been imported from Britain, particularly London), and it created a market within Continental Europe for surveying and astronomical instruments. During the first two decades of the nineteenth century, reform-oriented German states were forming a network of meridian astronomers:

Wilhelm Olbers, Friedrich Wilhelm Bessel, Heinrich Christian Schumacher, and Carl Friedrich Gauss. All these astronomers were also undertaking ordnance surveys of their lands. Hence, there was an expanding market for high-precision astronomical telescopes and ordnance surveying instruments, as the blockade stimulated local production. Utzschneider's management, Fraunhofer's lenses, and Reichenbach's instruments cornered that market. Indeed, by 1825, Fraunhofer's Optical Institute had become a purveyor of optical equipment to a vast European market.[46]

The Bavarians faced the lens crisis occasioned by the blockade of 1806 by manufacturing their own. Niggl, the optician of the ordnance surveying project, was hired by Utzschneider to be the optician of the Optical Institute, a part of the Mathematisch-mechanische Institut. Niggl, however, was unsuccessful at producing the glass needed to construct achromatic lenses, and was forced to leave the institute in 1807. Utzschneider scoured the monasteries of southern Bavaria, Nuremberg, the Bavarian Forest, northern Bohemia, and the Tirol for glassmakers until he learned of a Swiss watch and bell maker in the Prussian-ruled Swiss canton of Neuchâtel, in the city of Les Brenets, named Pierre Louis Guinand.[47] Utzschneider headed for Switzerland, where Guinand was preparing a lengthy manuscript for Utzschneider, entitled "Memorandum on Making Glass, especially of High Refraction for the Manufacture of Achromatic Telescopes."[48] Utzschneider was impressed with Guinand's glass samples, which were free of bubbles and striae, and decided to hire him.[49] Guinand traveled to Bavaria with his wife, Rosalie, and signed a contract with Utzschneider to direct the production of glass for achromatic lenses at the Optical Institute in May of 1806.[50]

A year earlier it had become clear to Utzschneider that the Mathematisch-mechanische Institut needed a large working space to manufacture optical glass in addition to the area needed for the manufacture of optical instruments. In that same year, using his previous governmental links, he purchased the secularized Benedictine monastery at Benediktbeuern, approximately 55 kilometers south of Munich, in the foothills of the Bavarian Alps. That space was to become the site of achromatic-lens production. For skilled labor, Utzschneider directed his attention to the monasteries, renowned for glass production.[51] He also drew upon workers from a nearby quartz quarry and workers familiar with the forests surrounding Benediktbeuern. All these enterprises, as we shall see below, were crucial to the erection of two glass ovens for flint and crown glass, which were manned by Guinand and his young assistant, Fraunhofer.

Secularization

The Bavarian Secularization of 1803, which played a vital role in Fraunhofer's production of achromatic lenses, was the culmination of a series of events, started in the 1770s, which had shifted bureaucratic power from the church to the state in Roman Catholic Bavaria and Austria.[52] Throughout the eighteenth century, there had been calls from Catholic monarchs to dissolve the religious sovereignties and principalities (geistliche Fürstentümer) of the Holy Roman Empire. Since the tenth century, German archbishops and bishops had also been Reichsfürsten, or princes of the empire, who had used and abused the power provided them by the pope. Hence, there had been a clear connection between religious orders and the governing of the territory.[53]

Kaiser Joseph II of Austria had closed down approximately 800 monasteries during his ten-year reign. Revolutionary France followed suit. The National Assembly had decided in favor of secularization on 2 November 1789.[54] On 12 July 1790 the clergy's civil constitution had been enacted. By dismantling the traditional institutional form of the French Church and destroying its corporate status and autonomy, the constitution had made all clerics salaried civil servants of the French nation-state. The Church had been severed from Rome. After Napoleon had put a stop to the persecution of the clergy in 1801 and 1802, Pope Pius VII had finally recognized the expropriation of the church's wealth to the state, which had actually been carried out in 1790. As Chadwick has argued, "The Revolution stood, first, for the demolition of the old Gallican Church of France; second, for the kidnapping of the pope and confiscation of his inheritance; third, for atheism or rationalism."[55]

During the eighteenth century, 1000 years after its founding, the monastery at Benediktbeuern had reached the zenith of its cultural achievements. The monastery library, which housed more than 25,000 books and more than 1000 paintings, was considered one of the finest in Bavaria.[56] Several Benedictine abbots were renowned for their scientific acumen. Pater Anand Fritz of Tölz, abbot from 1784 until 1796, had been an accomplished investigator of nature, particularly physics and chemistry.[57] Many clergymen specialized in mathematics and physics, particularly optics. Benediktbeuern clergymen had served as professors in the physical sciences at the universities of Salzburg, Ingolstadt, and Landshut.[58] Benedictine and Amarian cloisters also possessed impressive physical, astronomical, and mathematical instrument collections that, after the secularization, became the property of the Royal Bavarian Academy of Sciences in Munich. Kloster St. Emmeram had a particularly impressive collection, including instruments made by G. F. Brander.[59]

The monasteries had been, in effect, microcommunities. The abbots and padres hired skilled workers (geschickte Handwerker) from the surrounding community. The monasteries had also been used by young journeymen (Gesellen) to acquire the necessary skills of a particular trade.[60] In Benediktbeuern, up until the secularization, 207 men and women had been employed, including 36 padres and Benedictine brothers and four mendicant monks.[61] The remaining 177 employees were workers not religiously affiliated with the Benedictine order. The skilled laborers of this community included a mechanic, a sawmill operator, two blacksmiths, nine "forest scientists/economists" (Forstwissenschaftler/wirtschaftler), and six brewers.[62] Hence, when Utzschneider purchased the former monastery in 1805, not only did he inherit a huge working space; he also had a rather large labor force upon which to draw for his Optical Institute.

The Bavarian Secularization, taking its cue from the French, resulted in the usurpation of the property of the bishops, including the monasteries and convents over which they presided. The French wished to build up the German middle and lower states, particularly Bavaria, in order to establish a buffer against Austria. The funds for this growth and strengthening came from the secularized monasteries and properties of the bishops. The official document announcing the secularization and the empowering of German rulers, the Reichsdeputationshauptanschluß of 25 February 1803, stated: "All goods of the funded convents, abbeys, and monasteries . . . will be placed completely and freely at the disposal of the respective rulers (Landesherren) for the purposes of paying the cost of church services, instruction, and other useful institutions as well as for the distribution of their finances."[63] This Reichsgesetz was the result of the workings of Montgelas and the aggression of Napoleon. In 1796 Montgelas had drawn up a plan of reform for Bavaria and his exiled prince, Maximilian Joseph IV. Montgelas had plotted a scheme whereby monasteries would be secularized and the total number of monks would be reduced. The resulting funds would be at Maximilian Joseph IV's disposal to fuel a stagnant economy and reimburse Bavaria's debtors. A similar program had been initiated by Karl Theodor, independent of Montgelas and Maximilian Joseph IV, in 1799, when he had levied a tax of 15 million guilders on the religious institutions of Bavaria during the nation's financial emergency.[64] The institutions were forced to compromise their monasteries in order to satisfy Napoleon's plans for Bavaria. According to a diplomatic deal struck in Paris in January of 1801, 70 monasteries and convents in Altbayern were to be disbanded and the resulting property and funds given to Maximilian Joseph IV.[65] Montgelas's official recommendation to the prince came on 10 September 1801.[66] First, the monasteries of the mendicant order were to be secularized since they were not protected by the Vatican. Second, he wished to disband fourteen abbeys, which were indeed protected by the Vatican. At first Maximilian Joseph IV did not follow Montgelas's recommendation, fearing that this would only augment the civil unrest throughout the region. Instead, the prince ordered the complete inventory of all monasteries and abbeys of the region, and only reluctantly followed through with the secularization two years later.

Although the secularization provided the Bavarian government with a large source of income, an experimental space, and skilled labor, the abbots—not surprisingly—fought furiously to keep their monasteries, and their interests, intact. Two abbots in particular were quite outspoken: Ruppert Kornmann of Prüfening and Karl Klocker of Benediktbeuern. They argued that the finances raised were not sufficient to solve Bavaria's financial problems. They also claimed that the government was destroying an educational, cultural, and scientific institution. They were concerned with the future of the monasteries' rich collections of books, natural history cabinets, paintings, antiques, and mathematical and physical instruments, and they objected rather strongly to their sacred labor practices' being exploited for an intruding business.[67] Benedictine monasteries had a long tradition of scientific and artistic achievement that played a vital role in the cultural history of Bavaria. But despite the protests of Kornmann and Klocker, and despite the fear of civil unrest, secularization was inevitable.

The Content of Fraunhofer's Optical Work

It was in 1806, recall, that Utzschneider intervened in Fraunhofer's apprenticeship with the glass cutter Weichselberger and appointed Schiegg and Soldner to instruct the boy in physics, optics, and mathematics.[68]

Utzschneider provided Fraunhofer with the necessary textbooks on optics by Abraham Gotthelf Kästner, Georg Simon Klügel, and Joseph Priestley, and he studied the books, as far as can be historically reconstructed, with Utzschneider, Schiegg, and Soldner.[69] Fraunhofer began his apprenticeship in Munich; eight months later, at the beginning of 1807, he was transferred to the Optical Institute in Benediktbeuern. Fraunhofer thereby became the journeyman to Niggl, the optician of both the ordnance surveying project and the Optical Institute, with whom Fraunhofer had studied on Sundays from 1801 to 1804. It was Fraunhofer's job to learn how to produce achromatic lenses for theodolites, telescopes, and repeating circles for use by the Bureau topographique and the Bureau de cadastre.

The young Fraunhofer showed immense promise. After astonishing Niggl and Utzschneider with his manual skills, he quickly began to apply Schiegg's and Soldner's theoretical training in geometrical optics. In 1807 he wrote an essay on catoptrics dealing with the deviation of points outside of the optic axis of telescope mirrors,[70] demonstrating that hyperbolic mirrors produced clearer images than parabolic mirrors in the construction of reflecting telescopes, and describing his newly invented machine that could cut segments of hyperbolic mirrors.[71] Fraunhofer expanded his work to encompass polishing lenses and invented a polishing machine that would improve upon the spherical form of an objective lens produced by cutting.[72] Such a machine was necessary to prevent spherical aberration.

Of particular importance to this essay was Klügel's *Analytische Dioptrik* of 1778, which Fraunhofer later in life was to inform Carl August Steinheil was the only book that a Praktiker of optics needed.[73] Klügel, a professor of mathematics at the University of Helmstedt and later at Halle, authored various books and articles on geometrical optics for opticians and instrument makers. Klügel's text, which was dedicated to Euler, was divided into two parts. The first offered a general theory of telescopes and instruments of magnification; the second detailed the application of the general theory to optical instruments, including refractors, reflectors, and microscopes. The first part of Klügel's tome included sections on calculating the refraction of rays due to unequal refrangibility, the deviation of rays that originate from a point outside the optic axis, and the deviation of rays due to the shape of optical glass. The second part offered both methods of calculating the relationship between the refractive and dispersive powers and suggestions on the construction of object lenses free of chromatic and spherical aberration. His geometric and trigonometric calculations were typical of those used by opticians in the late eighteenth century: rather elementary, generally quadratic and cubic trigonometric equations, with differentials used sparingly.

Fraunhofer's early essay on catoptrics drew upon Klügel's calculations. For example, the young optician used Klügel's calculation for the derivation of light rays that originate outside the optic axis and Klügel's equation for the deviation of a light ray through a single spherical mirror infinitely removed form the object being viewed.[74] But Fraunhofer did not follow Klügel's text uncritically; he expanded on it considerably. For example, he deployed differential equations for a second-order calculation of the deviation of the ray outside the optic axis. Indeed, although he drew on Klügel's work, his mastery was evident from his derivation of the main ray of light. Rather than assume its trajectory through the surface apex of the mirror, as Klügel and others of the period had, Fraunhofer calculated the main ray through a point in front of the mirror's surface apex corresponding to the point where the ray cuts across the optic axis.[75] From an early age, then, Fraunhofer possessed an impressive range of analytical tools applicable to geometrical optics. Although he never attended a university, or even a Gymnasium, the training he received from Utzschneider, Soldner, and Schiegg, combined with his understanding of optical texts such as Klügel's, seemed more than adequate. His practical knowledge, combined with his newly acquired theoretical knowledge of geometrical optics, greatly impressed Utzschneider. On 4 February 1809, Utzschneider wrote to Fraunhofer informing him that he wanted the young apprentice to turn his attention to dioptrics rather than catoptrics.[76] By 1809, Fraunhofer was assisting Guinand in the manufacture of achromatic lenses for telescopes and microscopes.[77]

After signing a contract with Utzschneider and Reichenbach on 7 February 1809, Fraunhofer wanted to direct all aspects of optical glass manufacture.[78] In September of 1811, much to Guinand's dismay, Utzschneider granted permission to Fraunhofer to direct the production of optical glass at Benediktbeuern. He was then in charge of 48 people at Benediktbeuern, including twenty polishers, two tube drawers, one heater, five turners, one glass pourer, one assistant optician, and numerous skilled artisans, many of whom were young lads from the village of Benediktbeuern.[79] Bitter disputes erupted between the old Swiss bellmaker and the young, impatient

Bavarian. These disputes were never resolved, and Guinand left Benediktbeuern to return to his home in Les Brenets on 20 December 1813.[80] Less than two months later, Reichenbach decided to direct his attention away from the Optical Institute and concentrate his efforts on the Mathematisch-mechanische Institut in Munich. On 7 February 1814, the official contract among Reichenbach, Utzschneider, and Fraunhofer was dissolved and Reichenbach left the firm. A new contract was established between Fraunhofer and Utzschneider on 20 February 1814.[81] Fraunhofer thereby became the sole director of glassmakers at the Optical Institute. The correspondence between Utzschneider and Fraunhofer suggests that Fraunhofer had already been orchestrating the management of the institute for several years.[82] After the director of the mechanical section of the Optical Institute, Sigismund Rudolph Blochmann, left Benediktbeuern in 1818, Fraunhofer became director of all sections of the Optical Institute.[83] As a result of his superior method of optical glass production, Fraunhofer enjoyed a lifestyle far more lavish than most skilled artisans by virtue of his optical training.[84]

Production of Achromatic Glass

Optical glass refracts the colored rays of light from a distant object and (ideally) causes those rays to converge to a single point. This condition can be met only if the glass is of uniform density throughout. The ingredients of the optical glass vary, depending on whether flint or crown glass is being manufactured. The ingredients must be melted at a sufficiently high temperature to ensure liquidity, thereby allowing the bubbles, which are formed throughout by heating, to rise to the surface and escape. The batch of glass must be stirred throughout the lengthy process in order to guarantee homogeneity. The molten glass is then cooled to room temperature. The resulting glass blank can then be cut, ground, and polished.

Fraunhofer's recipes and the specific procedures he employed in making optical glass were never published, owing to considerations of secrecy. Indeed, not much was ever written down concerning the actual skills and practices involved. Fraunhofer taught his apprentices (unskilled boys from the surrounding community) and his future master glassmen by word of mouth and personal demonstration.[85] Fraunhofer's instruction mirrored the practice of Bavarian guild masters. Hence, a total reconstruction of his practices remains an impossible task for the historian. His glass recipes, however, have been recovered, after remaining a Bavarian state secret for nearly

Figure 3.1
A Fraunhoferian flint-glass prism. *(Deutsches Museum, Munich)*

100 years.[86] His flint glass was composed of quartz from the Tirol, potash (potassium carbonate, K_2CO_3), saltpeter (potassium nitrate, K_2NO_3), and red lead (Mennige, Pb_3O_4).[87] He varied the ratios of the ingredients in a trial-and-error fashion until samples with the desired dispersive and refractive indices were produced. Flint glass samples that required high dispersive and refractive indices had a high percentage of red lead, while samples requiring low dispersive and refractive indices contained a low ratio of red lead. His crown glass contained silica from sand, calcium carbonate, and potassium carbonate in the form of potash.[88] Again, he varied the ratios of these ingredients, testing literally hundreds of permutations of the original recipe. Guinand and Utzschneider had two different ovens built in

Figure 3.2
Fraunhofer's optical-glass probes, c. 1820. *(Deutsches Museum, Munich)*

Benediktbeuern: one for flint glass, which was larger since the temperature to melt the red lead needed to be higher, and one for crown glass. Wood heated the glass ovens, rather than coal, which was used in Britain. Since wood burns at a much lower temperature than coal, Fraunhofer and Guinand needed to solve problems inherent to glassmaking without the benefit of high temperatures.[89]

Good optical glass has five primary characteristics: homogeneity (uniformity in chemical composition), freedom from color, total transparency, a high degree of chemical and physical stability, and precisely known refractive indices for different colors (which Fraunhofer indeed determined). Glass that is physically homogeneous is free from striae, streaks, and bubbles. Striae generally possess a lower refractive index, thereby deflecting the paths of transmitted light rays and resulting in an impairment of the image. The slightest difference in the refractive index (even in the fifth decimal place—striae can affect the third decimal place) can be detected on a warm day, when rising hot air currents render distant objects unclear. In high-power telescopes, the rays converge to an image point under a small angle,

Figure 3.3
Fraunhofer's glass hut in Benediktbeuern (c. 1820), as depicted by R. Wimmer. *(Deutsches Museum, Munich)*

and even the slightest deviation in path drastically alters the quality of the image.

During all stages of melting, volatile matter escapes from molten glass. If this matter cools before reaching the surface, it results in a bubble. Also, oxygen gas is released from the litharge (PbO) or red lead, and CO_2 in the case of potassium carbonate. If the gas cools before it escapes from the surface, bubbles will form. Bubbles can be deleterious to optical glass—particularly if they are formed in the image plane of a telescopic system, since they can disturb details in the field of view. They can lift a heavier layer of the molten glass into a lighter and less refractive layer.

It proves to be rather difficult to alleviate striae, streaks, and bubbles. Intense stirring can eliminate striae and streaks, but it results in more bubbles. Achieving homogeneous glass is, therefore, a trial-and-error procedure. Even Fraunhofer could not always guarantee a piece of homogeneous glass. We often read in what few notes he left behind that bubbles were found in his samples. The more he practiced his craft, the more likely he would be successful, but no glassmaker could boast a flawless record. Homogeneous flint glass was much more difficult to manufacture than homogeneous crown glass for two reasons. First, one of the main ingredients of flint glass is lead oxide (in the form of red lead or litharge), a very dense material. It sinks to the bottom of a glass pot resulting in a density gradient, which will give rise to striae if the glass is not stirred thoroughly enough. Second, litharge and red lead are very caustic, particularly under extreme heat. They can destroy the crucibles or pots that hold the molten glass. The glass then fuses to the furnace's floor and walls.

Guinand invented and Fraunhofer improved upon a stirring technique that created homogeneous glass blanks without bubbles or striae. Their technique—necessary because homogeneity is more difficult to achieve at the low temperatures reached when wood rather than coal is used as a fuel, utilized a wooden stirring rod that was protected from the heat by an earthenware-clay coating. This type of stirrer is similar to those used by bellmakers for the mixing of various metals in the liquid state. Guinand, who had been trained as a bellmaker, would have most likely have learned such a technique as a young apprentice. Metal rods would melt and contaminate the mixture. The stirrer (Tonrührer), which Fraunhofer described as being 6 Parisian inches wide and 27 Parisian inches long (approximately 6½ English inches by 28⅘ English inches),[90] was carefully heated until bright red and then placed on the wall of the melting furnace until it turned white. It was slowly inserted into the molten glass and then withdrawn and permitted to remain for an hour on the edge of the glass pot, floating on top of the molten glass and preventing the formation of harmful bubbles. This allowed the glass to penetrate the clay coating and drive out any gases that would otherwise contaminate the molten glass. The rod was then reinserted and stirred by a mechanical device.[91] Guinand and Fraunhofer's technique of stirring, combined with Fraunhofer's superior polishing technique and very pure quartz obtained from Zillertal in the Tirol, all resulted in the superiority of the Optical Institute's lenses. Fraunhofer's first major contribution to optics was his manufacture of achromatic glass and its conversion to optical lenses. Fraunhofer then needed only a precise, replicable technique that could calibrate the accuracy of his lenses.

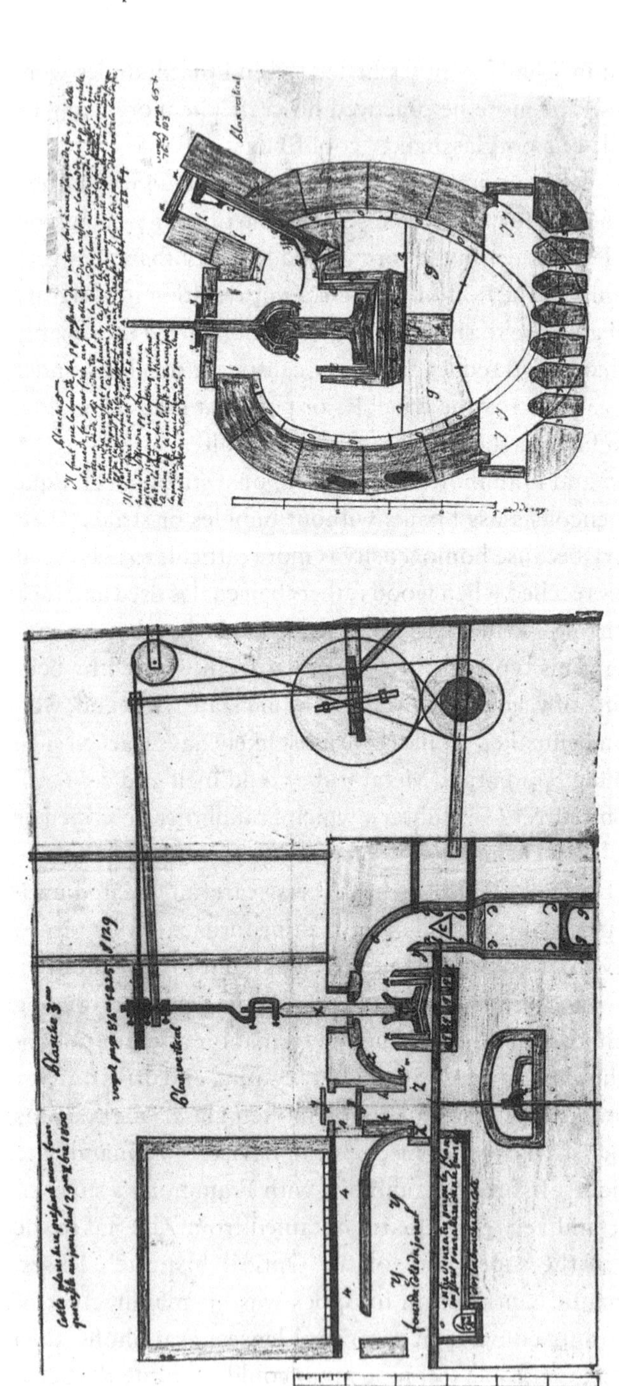

Figure 3.4
Two glass ovens with stirring apparatus (c. 1805–06), as sketched by P. L. Guinand. *(Deutsches Museum, Munich)*

Figure 3.5
The melting and glowing oven at Benediktbeuern, c. 1815. *(Deutsches Museum, Munich)*

The Six Lamps Experiment and the Solar Light Experiment

Fraunhofer's essay Bestimmung des Brechungs- und Farbenzerstreuungs-Vermögens verschiedener Glasarten in bezug auf die Vervollkommnung achromatischer Fernröhre (Determination of the Refractive and Dispersive Indices of Differing Types of Glass in Relation to the Perfection of Achromatic Telescopes), written in 1814–15, was the culmination of his experiments on perfecting the manufacture of achromatic lenses.[92] It was certainly not an attempt to explain theoretically the nature of the solar dark lines (later to be called absorption or Fraunhofer lines) or the lamp lines (later to be called emission lines). Its sole purpose was to publish Fraunhofer's method for improving the construction of achromatic lenses for telescopes. Indeed, the essay commences by claiming that

> the calculation of an achromatic object glass, and generally that of every achromatic telescope, necessitates a precise knowledge of the ratio of the sines of incidence and refraction [Snel's Law], and of the ratio of various types of glass which are used in the construction of telescopes . . . Experiments repeated during many years have led me to discover new methods of obtaining these ratios, and I have therefore obeyed the wishes of several scholars [astronomers and experimental natural philosophers] in publishing these experiments, in the order I made them,

with the necessary modifications that the experiments themselves forced me to introduce.[93]

Fraunhofer concludes the piece with this admission:

In making the experiments of which I have spoken in this memoir, I have considered principally their relations to practical optics. My leisure did not permit me to make any others, or to extend them any farther. The path that I have taken in this memoir has furnished interesting results in physical optics, and it is therefore greatly hoped that skillful investigators of nature would condescend to give them some attention.[94]

Ever since Isaac Newton's work on the spectrum during the late seventeenth and the early eighteenth century, experimental natural philosophers and opticians had attempted in vain to determine the refraction of each colored ray. But since colors of the spectrum seem to be continuous, no precise methods could be established for choosing which colors to measure. Fraunhofer himself admitted in his essay that "it would be of great importance to determine for every species of glass the dispersion of each separately colored ray. But since the different colors of the spectrum do not present any precise limits, the spectrum cannot be used for such."[95]

At first, Fraunhofer, like many others before him, attempted to circumvent this problem by focusing his attention on colored glasses and prisms filled with colored fluids. He hoped to determine the refractive index of a glass sample for the color of light supplied by these filters, which, he hoped, would permit only homogeneous light to pass. Despite various attempts to produce such a glass or fluid, the emergent light never proved to be truly monochromatic; a mixture of spectral colors always resulted. Fraunhofer also attempted to use colored flames produced by burning alcohol and sulfur. But these flames, too, produced a spectrum when viewed through one of his prisms.[96] He did, however, notice during these investigations that the spectra produced by the alcohol and sulfur flames were marked by a clearly defined line in the orange region. This line (later discovered to be two lines very close to each other and called the D-sodium couplet) proved to be crucial to Fraunhofer's subsequent research.

After giving up on colored glasses and liquid-filled prisms, Fraunhofer decided that he wanted to view the spectrum produced when the light of a burning lamp[97] was refracted by a prism and then viewed through a telescope mounted on a modified theodolite (an ordnance surveying instrument originally designed to measure angles for the production of maps). The

Figure 3.6
Fraunhofer's prism spectral apparatus, c. 1815. *(Deutsches Museum, Munich)*

modified theodolite could measure the angle of emergence from the subject prism for each colored ray. Unfortunately, the rays of light falling onto the subject prism would not be parallel, so that the angle of incidence would not be the same for each one, rendering the modified theodolite's measurement useless. In order to ensure that the rays striking the subject prism would be parallel, Fraunhofer substantially increased the distance between the lamp and the prism. But, he noted, although the rays now all had measurably the same angle of incidence, the increased distance resulted in some of the refracted rays' missing the prism altogether, producing an incomplete spectrum. To ensure that rays incident on the subject prism remained parallel, that an entire spectrum would be generated, and that the light would be intense enough to be seen through minute slits at such a large distance, Fraunhofer used six lamps.

Fraunhofer placed the six lamps behind a shutter (BC in figure 3.8) 1.5 Bavarian inches high and 0.007 Bavarian inch thick. The shutter was pierced by six narrow slits, each slightly less than 1.5 inches high and 0.05

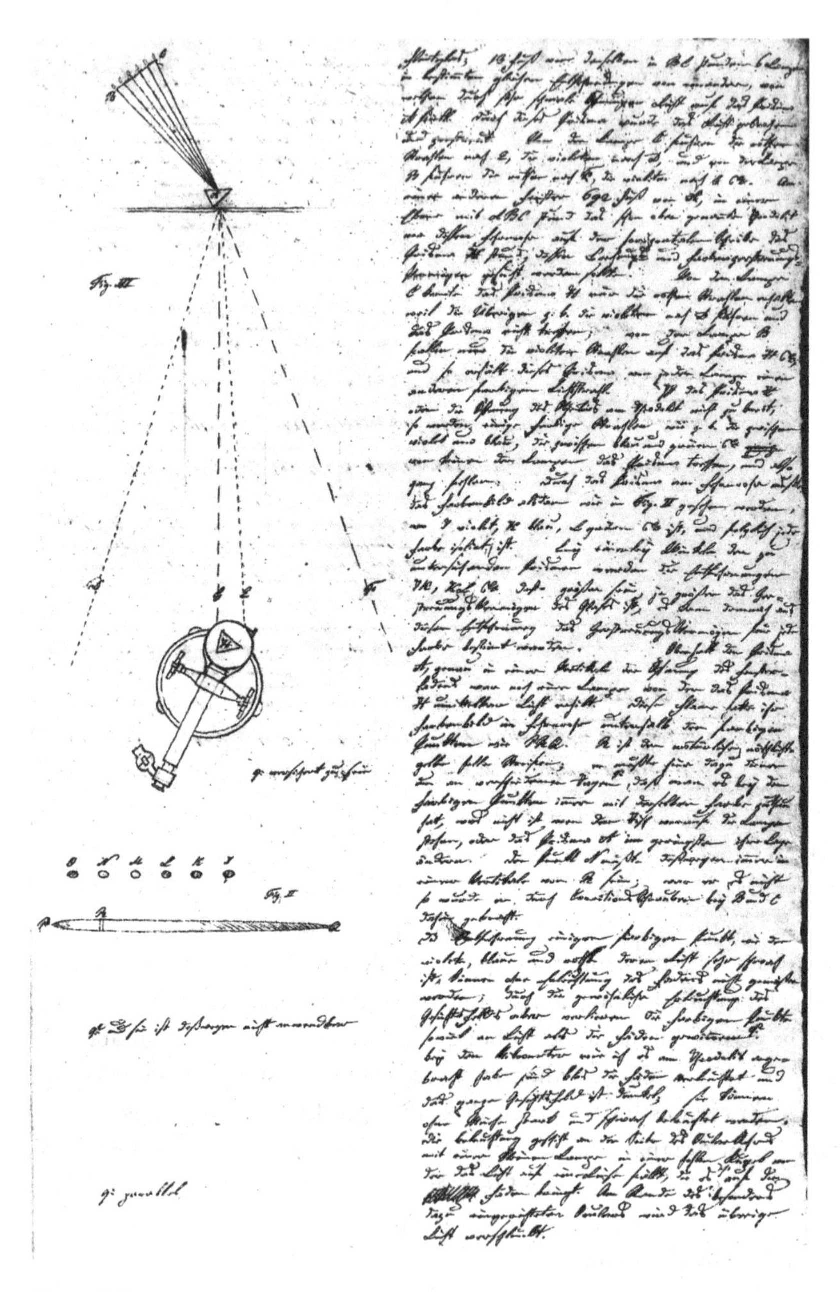

Figure 3.7
A page from Fraunhofer's manuscript "Determination of the Refractive and Dispersive Indices of Differing Types of Glass." *(Deutsches Museum, Munich)*

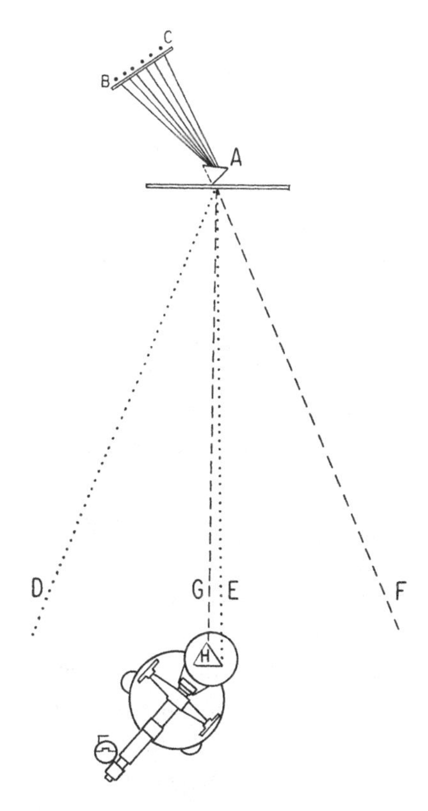

Figure 3.8
The design of Fraunhofer's Six Lamps Experiment.

inch wide. The six lamps were placed 0.58 inch apart, directly behind a shutter, with each lamp centrally located behind one of the slits. The shutter was placed 13 Bavarian feet (slightly more than 4.2 meters) from prism A, which was made of flint glass, and at an angle of approximately 40°. The light, having passed through the slits, was now refracted by the prism and decomposed into colors. The dispersed light then traveled through a second slit placed directly behind the prism, which accordingly blocked a portion of the emergent beam. Some of the rays were channeled to the site of a theodolite located in Fraunhofer's laboratory at the very great distance of 692 Bavarian feet (approximately 225 meters) from the six lamps.

The six-shutter mechanism controlled the angles at which light from each lamp struck the surface of prism A, thereby determining the locus of the corresponding spectrum. For example, from the lamp at C, red rays refracted to E and violet rays to D. From lamp B the red rays traveled toward F and the violet rays toward G. On the theodolite Fraunhofer placed a prism H whose index of refraction for the different colored rays was to be determined. He then adjusted the distances of the six-shutter mechanism from prism A, of A from the single shutter, and of the single shutter from prism H in such a manner that prism H received only red rays from lamp C and only violet rays from lamp B. The intermittent lamps supplied the other colors of the spectrum.[98] The spectrum of rays passing through the small aperture below A and then through prism H will appear in the modified theodolite's telescope as depicted in figure 3.7, where I is the violet, K the blue, L the green, and so on; each spectral color will appear at a unique locus. Fraunhofer ground down the angle of prism H until all the rays from the six lamps sensibly emerged from H at a single point (though, of course, each colored ray exited from that point at a unique angle with respect to the face normal there). The object lens of the modified theodolite's telescope was aimed at that point, thereby enabling Fraunhofer to see the entire spectrum and measure each color's dispersion. The distances *ON*, *NM*, etc. (figure 3.7) increased with dispersive power under these conditions. Since these distances and the incident angle could be measured by the modified theodolite to the nearest arc second, Fraunhofer could now determine the index of refraction to six decimal places for each colored ray for each type of refracting substance.[99]

To see whether other sources of light produced the same sort of lines as the sodium lamps had, Fraunhofer decided to use the sun as his source. He placed his modified theodolite and prism in a darkened room with a window that was covered by a shutter. He cut a vertical slit in the window shutter 15 arc seconds wide and 36 arc minutes high (approximately 0.6 mm wide by 80 mm high) with respect to the center of the theodolite, allowing the solar rays to fall on a flint glass prism with an angle of 60° mounted on the theodolite 24 feet from the window. The prism was placed in front of the telescope's objective lens in such a manner as to ensure symmetric passage[100] (because, as Newton had argued 100 years earlier, that position minimizes the effect of errors in setting the incidence). Fraunhofer remarked: "In looking at this spectrum for the bright [sodium] line, which I had discovered in a spectrum of artificial light, I discovered instead an infinite number of vertical lines of different thicknesses. These lines are darker than the rest of the spectrum, and some of them appear entirely black."[101] Fiddling

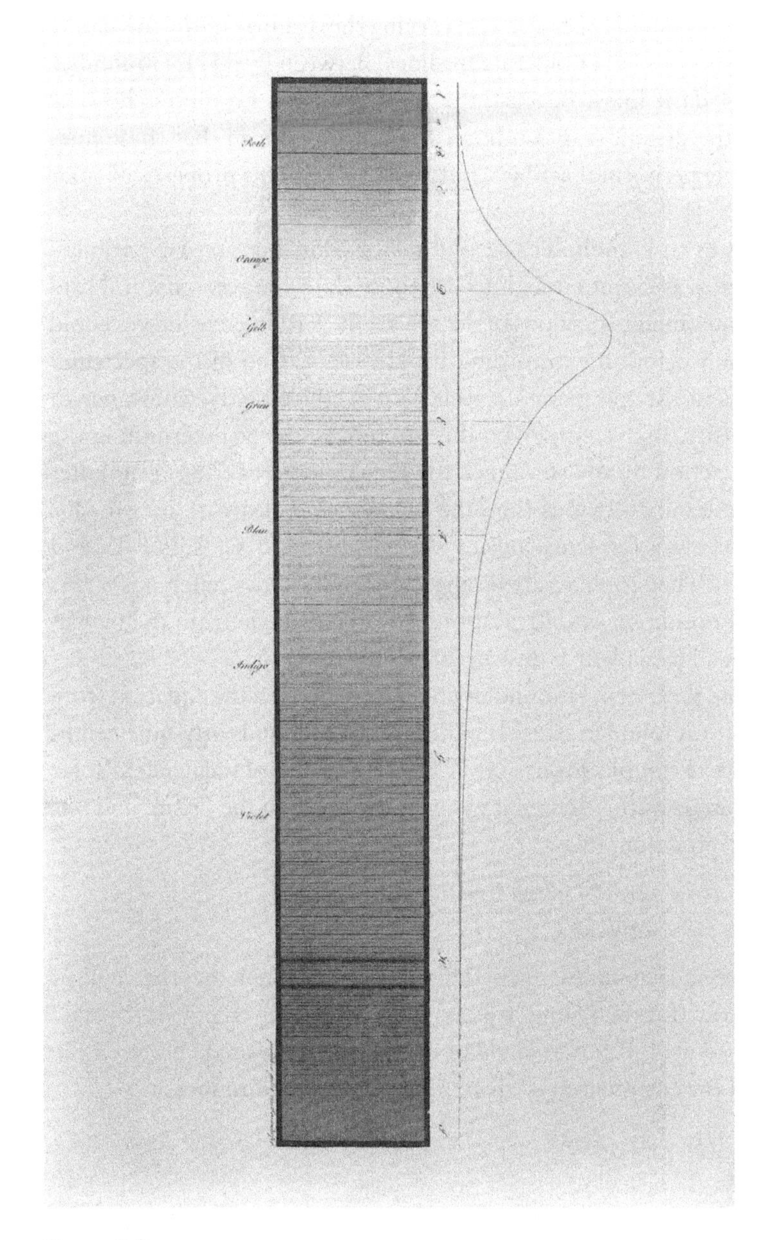

Figure 3.9
The dark lines of the solar spectrum: a black-and-white copy of the color plate as depicted by Fraunhofer. *(Deutsches Museum, Munich)*

with the window-shade aperture and varying the distance of the theodolite from the window did not obliterate the lines. Between B and H, Fraunhofer counted 574 dark lines. Because the lines persisted no matter how he rearranged the distances, Fraunhofer became convinced that these lines were not an experimental artifact, but were an inherent property of solar light.

This leads us to Fraunhofer's second major contribution to experimental optics. He was clever enough to use those dark lines as a natural grid that demarcates minute portions of the spectrum.[102] Refractive indices could now be obtained for an extraordinarily precise portion of the spectrum: "As the lines of the spectrum are seen with every refractive substance of uniform density, I have employed this circumstance for determining the index of refraction of any substance for each colored ray."[103] Fraunhofer then chose the most obvious (i.e., the thickest and clearest) lines for his determination of the refractive indices of a glass prism: *B*, *C*, *D*, *E*, *F*, *G*, and *H* (figure 3.9). They could easily be aligned with the cross-hatchings on his theodolite. Fraunhofer would simply read off the angle from the instrument's vernier. He made five measurements for each line.

To compute the index, Fraunhofer could have derived the equation from similar equations found in several eighteenth-century and early-nineteenth-century optics textbooks for prisms.[104] If σ is the angle of the incident solar ray, ρ is the angle of the emergent ray, ψ is the angle of the prism, and n is the index of refraction, then[105]

$$n = \frac{\sqrt{(\sin\rho + \cos\psi \sin\sigma)^2 + (\sin\psi \sin\sigma)^2}}{\sin\psi}.$$

As we have seen, Fraunhofer devised his experiments such that the angle of the incident ray (for the *D* line) was equal to that of the emergent ray (i.e., symmetric passage). If μ (the angle of deviation) is the angle between the incident and the emergent rays, then, under these circumstances,

$$n = \frac{\sin\frac{1}{2}(\mu + \psi)}{\sin\frac{1}{2}\psi}.$$

The angle μ was measured by the modified theodolite, as were the arcs *BC*, *CD*, *DE*, *EF*, *FG*, and *GH*. If n_E is the refractive index for the ray *E*, then

$$n_E = \frac{\sin\frac{1}{2}(\mu + \psi + DE)}{\sin\frac{1}{2}\psi},$$

$$n_F = \frac{\sin\frac{1}{2}(\mu + \psi + DE + EF)}{\sin\frac{1}{2}\psi},$$

$$n_G = \frac{\sin\frac{1}{2}(\mu + \psi + DE + EF + FG)}{\sin\frac{1}{2}\psi},$$

and so on.

Fraunhofer created scores of tables listing the refractive indices of the rays (each ray corresponding to a line) for different substances—flint glass, crown glass, oil of turpentine, and water, to name just a few. He then created tables of indices for combinations of refracting media in order to determine the combination that would correct chromatic aberration for the red and violet rays of the spectrum.

Fraunhofer had now provided opticians and experimental natural philosophers with a vastly more precise method for determining the refractive indices of glass samples than had ever before been attained. Previously, Fraunhofer himself had determined the relative dispersive and refractive indices of two kinds of glass by cementing them together, forming a single prism. If the two spectra produced by this compound prism appeared at the same place, without any reciprocal displacement, he concluded that their dispersive and refractive powers were the same and equal to the arithmetic average of the two extreme rays: red and violet. After the discovery of the lines, however, he quickly realized that two pieces of glass, which appeared to have the same refrangibility when employing the early method of testing, could actually have slightly different powers, as revealed by the existence in the overlap region of two sets of lines where there should be only one.

Fraunhofer had accomplished what Clairaut had attempted some 50 years earlier, namely to destroy the secondary spectrum resulting from the different partial dispersions of crown and flint glass, thanks to the Bavarian's glass. Because Fraunhofer's glass was of such superior quality, a new method of measurement could be introduced for the construction of achromatic lenses; one could now determine the refractive index for a very specific point of the spectrum. Clairaut had been unable to do that because of the inferior quality of French optical glass, particularly back in the 1760s. Indeed, so sensitive was Fraunhofer's new method that a difference in refracting power was found not only in different types of glass or between samples taken from different levels of the same melting pot or crucible of glass, but even between pieces taken from the opposite ends (top and bottom) of the same piece of a glass blank. Fraunhofer and Guinand next invented a method of stirring so efficient that, in a pot containing more than 400 pounds of flint glass, a specimen taken from the top and one from the bottom had exactly the same refractive index. Overcoming heterogeneity was not a problem for Fraunhofer, but (as we shall see) it was a major problem for Michael Faraday and for the British attempts to reverse engineer Fraunhofer's optical glass.

The dark lines of the solar spectrum, then, provided Fraunhofer with a tool for gauging the efficacy of achromatic-lens production. If the refractive indices determined by aligning the dark lines with Reichenbach's modified theodolite indicated that the glass was not suitable for constructing an achromatic lens (i.e., if another lens could not correct the first lens's chromatic aberration efficiently enough), Fraunhofer and his apprentices would alter the recipe by adding more or less lead oxide in the form of red lead, or by increasing or decreasing the time of stirring and cooling, or by changing how the glass blanks were cut and ground into lenses. Hence, Fraunhofer's measurement of the refractive and dispersive indices was connected to the production procedures in an unprecendented way. This portion of the process was an example of pröbeln—a Bavarian term similar to the German word probieren, meaning "to test" or "to try out." Pröbeln, inherently, cannot be precisely specified. It is a type of knowledge that by its nature does not lend itself to replication. It is part and parcel of artisanal knowledge, since only a skilled artisan knows when a trial has been successful.[106] Through this trial-and-error procedure, tested time and again by his technique of measuring solar lines, Fraunhofer produced the world's most coveted achromatic lenses.

The Practice of Secrecy: Public and Private Knowledge at the Monastery of Benediktbeuern

Because the Optical Institute was a business venture, Fraunhofer made it a practice to differentiate between public and private knowledge. On the one hand, he wanted to increase the visibility of the institute in order to increase sales. Hence, he needed to publicize aspects of his enterprise. This he accomplished by publishing lists of the Optical Institute's products as well as by entertaining experimental natural philosophers and demonstrating to them his method for the calibration of achromatic lenses.

Fraunhofer also wished to obtain credit for his work from the scientific community. This he was able to do by publishing his essay on the experiments that enabled such a calibration in the *Denkschriften der königlichen Akademie der Wissenschaften zu München*. On the other hand, certain forms of his artisanal knowledge had to be kept secret. Hence, Fraunhofer policed both the written word and the social space. He published neither the procedures nor the recipes for producing achromatic lenses, nor did he ever permit the experimental natural philosophers to witness the skills involved in lens production.[107]

The reputation of Fraunhofer's legendary workmanship quickly spread throughout the German territories, as ordnance surveying projects were being conducted in Prussia, Saxony, Lower Saxony, and Schleswig-Holstein during this period. Fraunhofer's workmanship was matched only by Utzschneider's entrepreneurial genius. Utzschneider published the Optical Institute's price lists for achromatic telescopes, theodolites, repeating circles, transit instruments, and optical lenses in journals affiliated with ordnance surveying and astronomy. These lists appeared, for example, in Franz Xavier von Zach's *Monatliche Correspondenz zur Beförderung der Erd- und Himmelskunde*, in Bernhard August von Lindenau and J. G. F. von Bohnenberger's *Zeitschrift für Astronomie und verwandte Wissenschaften*,[108] in J. C. Poggendorff's *Annalen der Physik und Chemie*, and even in optical texts such as J. J. Prechtl's *Praktische Dioptrik*.[109] Astronomers, ordnance surveyors, and mapmakers across Europe became familiar with the Optical Institute's products.

Experimental natural philosophers from Britain, France, and the German territories would visit the Optical Institute to inquire whether Fraunhofer could manufacture lenses to their specifications. For example, Gauss traveled to Benediktbeuern in 1816 to place an order for astronomical instruments for his new observatory at the University of Göttingen. Other astronomers followed suit. Olbers from Bremen, Schumacher from Hamburg/Altona and Copenhagen, and Bessel from Königsberg all ordered their astronomical instruments from the Optical Institute. Fraunhofer's reputation was also enhanced by the scores of articles published by Bessel, Olbers, and Struve announcing the discovery of stars hitherto unseen but now made visible by Fraunhofer's refractors. And finally, of course, Fraunhofer's reputation soared as he continued to publish his research on light and physical optics in prestigious journals.

Visitors to Benediktbeuern were not restricted to astronomers and experimental natural philosophers. King Maximilian I; his son, King Ludwig I; Montgelas, the leader of Bavaria's reforms in the first two decades of the nineteenth century; and the king of Prussia all visited the secularized cloister to see the Optical Institute. Czar Alexander I of Russia had originally planned to pay a visit on his return from the Congress of Vienna, but canceled his trip at the last moment[110]; he did, however, place an order for achromatic telescopes for his Russian universities, including the renowned refractor at Dorpat.

Although there are numerous examples of distinction between public and private knowledge in Benediktbeuern, four are especially interesting. First, on 14 April 1810 the co-owner of the Optical Institute, Utzschneider, wrote to Fraunhofer: "Should the French, or some other foreigners now come to Benediktbeuern, do not show them the flint-glass oven. Also, be sure that they do not come in contact with Mr. Guinand. A certain Marcell de Seoren [?] has come for this reason. . . . One should show him nothing."[111] A second instance also involved Guinand, the glassmaker at Benediktbeuern who had taught Fraunhofer his trade. As a part of Guinand's contract of 20 February 1807, he received from Utzschneider an apartment for him and his wife and five cords of wood "on the condition that he never communicates the secret of the fabrication of flint and crown glass to anyone."[112] After numerous quarrels with Fraunhofer, Guinand left Benediktbeuern in December of 1813. Utzschneider agreed to pay him an annual pension as long as he and his wife promised not to divulge the secrets of manufacturing flint and crown glass or work for another optical institute that could rival Benediktbeuern.[113] Such deals were commonplace. Guinand agreed, but broke his promise shortly thereafter. The third example of secrecy involved Johann Salomo Christoph Schweigger, professor of physics and chemistry at the University of Erlangen (and later of Halle), who wished to visit Fraunhofer at the Optical Institute. Fraunhofer wrote Schweigger expressing his delight to entertain such a prestigious guest, but informed the experimental natural philosopher that although he would be glad to teach him how to use the dark lines of the solar spectrum to calibrate lenses, he would not be able to show the professor the glass hut or divulge any information on the manufacture of achromatic glass.[114]

The most informative example of secrecy involved Fraunhofer and John Herschel, Britain's leading experimental natural philosopher. On 19

September 1824, Herschel visited Benediktbeuern on behalf of the Joint Committee of the Board of Longitude and the Royal Society for the Improvement of Glass for Optical Purposes. British lensmakers and experimental natural philosophers had felt threatened by Fraunhofer's enterprise. Herschel hoped to take away information regarding Fraunhofer's glassmaking techniques, but was disappointed. Imagine Herschel's surprise when he—a representative of reformed, regal, industrial Britain, where glasshouses were located only in major cities—was greeted in a former monastery situated in the foothills of the Alps.[115] As will be discussed in chapter 5, Fraunhofer would not divulge to Herschel any of his practices for making achromatic glass.

Although, as we have just seen, visitors increase the visibility and fame of an enterprise, they can also be potentially dangerous, since they can witness methods of manufacture. Fraunhofer and Utzschneider therefore needed to switch from controlling only the written word to policing the social space as well. They controlled this space by prohibiting access to certain portions of their institute. Such restrictions could be executed rather efficiently, since Fraunhofer and Utzschneider could draw upon the Benedictine architecture, which was an instantiation of public and private monastic life.

Secrecy and Architecture of Benediktbeuern

> I swear further by the great God who has disclosed these things, to hand this book down to no one except to my son, when he has first judged his character and decided whether he can have a pious and just feeling about these things and can keep them secure. . . . Keep this a sacred thing, a secret not to be transmitted to anyone, and you will not as a prophet have given it away.
>
> —*Mappae clavicula* of the Benedictine Monastery at Reichenau[116]

In order to construct his lenses, Fraunhofer drew upon the architectural space and layout of a secularized Benedictine monastery—an architecture that instantiated three elements critical to the Rule of Saint Benedict: labor, silence, and secrecy.[117] A study of Fraunhofer can, therefore, offer an insight into the more general relationships between the scientific enterprise and architectural space.

I shall not claim that a Benedictine monastery was the only site where optical lenses could have been constructed. However, of all the abundant resources that Utzschneider, former Bavarian privy councilor, entrepreneur, and co-owner of the Optical Institute, had at his disposal, a Benedictine monastery best suited his needs for a glasshouse. First, Benediktbeuern was located in the midst of a large forest, where wood for fuel was in abundant supply. Second, a quarry of quartz, one of the key ingredients of glass, was situated in nearby Zillertal. Third, Benedictine monks and artisans from the surrounding communities were well versed in optical theory and practice and in the manufacture of glass.[118] For example, Pater Udalricus Riesch, a co-worker of Fraunhofer at the Optical Institute, had been a mapmaker at Benediktbeuern and possessed a thorough knowledge of surveying instruments.[119] Pater Josef Maria Wagner taught optics to the young apprentices working with Fraunhofer in the monastery.[120] Both Riesch and Wagner were Fraunhofer's closest friends.[121] Recall that Pater Ulrich Schiegg was a Benedictine monk, and Niggl was trained in optics by the Benedictines at Rott.[122] Indeed, Benedictine monasteries throughout Bavaria, Bohemia, and the Tirol had been sites of glass manufacturing for nearly 1000 years. Utzschneider, before hiring Guinand, had actually started his optical glass experiments in the Benedictine monastery at Ettal, where a labor force and glass ovens were still intact. Those experiments, however, failed to produce the quality of glass necessary for achromatic-lens production.[123] Skilled laborers also included the "forest scientists" (Forstwissenschaftler) of Benediktbeuern, who calculated the amount of lumber available in the surrounding forests for its use as fuel for glass melting. Although nontechnical glassmaking was normally restricted to common table glass, window glass, Benedictine brandy glasses, and reading lenses, skilled and unskilled artisans could provide assistance in glassmaking, cutting, and polishing lenses. But also needed, of course, were Guinand's and Fraunhofer's abilities to produce achromatic lenses, which are more difficult to produce than ordinary glass. Fourth, many monasteries and cloisters late in the eighteenth century and early in the nineteenth century possessed lavish collections of physical instruments and texts dealing with glass manufacture, optical theory, and astronomy.[124] For example, the Benedictine monastery of Kremsmünster in Austria had been a renowned astronomical observatory and center for ordnance surveying calculations since the 1770s.[125] Fifth, the monastery's large space enabled Fraunhofer to devise an experiment where the rays of light emitted from lamps were nearly parallel. This proved crucial for his use of those lines in the construction and calibration of achromatic lenses. Finally, Benedictine architecture reflected the public and private rituals associated with monastic life. Hence, Benediktbeuern could easily be transformed into an optical enterprise that required a demarcation between public and private access. All these factors made Benediktbeuern the ideal location for the Optical Institute.

Fraunhofer and Utzschneider's practice of silence and secrecy mirrored the centuries-old labor practices of the Benedictine monks. The labor of and the recipes for glassmaking, for example, were tightly kept secrets. For monks it was a vow of secrecy with God. Skilled laborers not belonging to any monastic order were notorious for their practice of guild secrecy. Technical knowledge, such as glassmaking, was the artisan's most prized possession. Economic considerations compelled artisans to keep their labor a secret from those who were not guild members.[126] By the late Middle Ages, as literacy spread, guild members had shifted from basing their teachings on oral instructions to writing down their technical secrets for future generations. This was particularly true of monastic craftsmen, whose literacy rate had been much higher than that of secular artisans.[127] These texts, however, were still not accessible to the public. Glassmakers often established symbols, much like alchemical symbols, in order to code their recipe books and explanations of glassmaking practices.[128] As Eamon has argued, these recipe books were "the record of trial-and-error experimentation. They [were] the accumulated experience of practitioners boiled down to a rule."[129] These recipe books and manuscripts were preserved in monasteries where such arts were practiced.[130] But it must be emphasized that the knowledge of artisans was most accurately transmitted by doing and by imitating and emulating the master, rather than by reading the printed word.[131]

Another extreme measure taken to ensure secrecy was the master glassmakers' decision rarely to disclose their last names for fear of being kidnaped by rulers of other regions.[132] Utzschneider and Fraunhofer's use of monks and local artisans for their sources of labor and optical knowledge was rather clever, since all their employees originated from cultures where secrecy was paramount. They were able to make use of existing monastic privacy and secrecy, even after the secularization of Bavarian monasteries in 1803, in order to create a scientific, technological, and private enterprise.

The monks' washroom at Benediktbeuern (A in figure 3.10) had been a public place where monks freely engaged in various discussions. It was

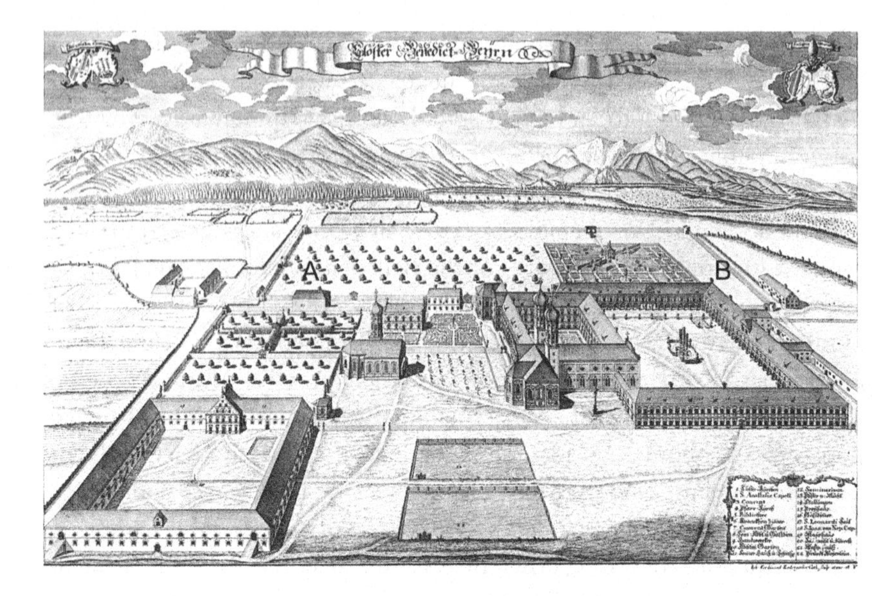

Figure 3.10
The Monastery at Benediktbeuern, c. 1800. *(Deutsches Museum, Munich)*

transformed into the achromatic-glass melting room. The glasshouse was off limits to visiting opticians and experimental natural philosophers, such as Herschel.

Entrance to Fraunhofer's laboratory (B in figure 3.10) was limited to those workers of Benediktbeuern who had optical expertise. The laboratory was built within the monks' cells, which were designed to reflect the importance of silence in the Rule of St. Benedict. Although it was therefore private, visiting opticians and experimental natural philosophers were taken there so Fraunhofer could demonstrate to them his technique of calibrating achromatic lenses. By showing visitors how he used the dark lines of the spectrum in producing achromatic lenses, rather than how the lenses were actually constructed, Fraunhofer ensured his institute's optical hegemony.

Monastic Labor, Architecture, and Space

Labor had always played a major role in Benedictine culture. The Benedictine monastery consisted of the school of divine service and the workshop where the monks labored at the artes spirituales; work was worship (laborare est orare).[133] The Rule of St. Benedict ordered monks to practice manual labor in any art that would serve for the benefit of the monastery, order, and humankind, and clearly explicates the importance of labor to the order: "Idleness is an enemy of the soul. Because this is so the brothers ought to be occupied at specified times in manual labor."[134] Craftsmen from the surrounding areas were requested "to ply their trade in the monastery."[135]

The importance of labor to Benedictine culture has been the object of much attention. Herbert Workman suggests that

> Benedict's success in linking Monasticism with labor was the first step in a long evolution. . . . whose main features demand attention. The change [from eremiticism to cenobitism] itself would have been of little value, at any rate viewed from the standpoint of social development, had it not been accompanied by the glorification and systematization of toil. With this addition the change lay at the root of all that was best and most progressive in Monasticism. Instead of the dervish of Eastern fancy, we have a colony of workers. Instead of the hermit crushed by the horror and loneliness of Nature in her most terrible aspects, we have the organized community . . . whose axes and spades cleared the densest of jungles, drained pestilent swamps, and by the alchemy of industry turned the sands into waving gold, and planted centres of culture in the hearts of forests.[136]

Both Max Weber and Lewis Mumford alluded to the importance of Benedictine notions of labor to the development of Western culture. Weber claimed that the Rule of St. Benedict

> had developed a systematic method of rational conduct with the purpose of overcoming the *status naturae*, to free man from the power of irrational impulses and his dependence on the world and on nature. It attempted to subject man to the supremacy of a purposeful will, to bring his actions under constant self-control with a careful consideration of their ethical consequences. Thus it trained the monk, objectively, as a worker in the service of the kingdom of God, and thereby further, subjectively, assured the salvation of his soul.[137]

Mumford asserted that Benedictine monasteries were the first example of a fusion of technology with moral force. Monasteries abandoned the master-slave relationship in economic organization of labor in favor of the "free man's moral commitment."[138] Monastic culture organized communities whereby groups structured themselves in "a daily routine of ordered activity" around manual and spiritual labor.[139] The Benedictine monastery "laid down a basis for order as strict as that which held together the early megamachines: the difference lay in its modern size, its voluntary constitution and the fact that its sternest discipline was self-imposed."[140] The Benedictine

Rule was successful in fusing economic and spiritual concerns precisely because it viewed the two as closely related.[141] This union formed the cornerstone upon which Benedictine culture was erected.[142]

Another important characteristic of Benedictine culture was its emphasis on silence and secrecy. Chapter 4 of the Rule of St. Benedict insisted that a Benedictine not be a murmurer or be fond of much talking.[143] Monks were to "practice silence at all times. . . . If any one shall be found breaking this rule of silence he shall be punished severely."[144] Silence performed two crucial functions in the lives of the Benedictines. First, it guaranteed self-discipline. Second, it ensured that the lifestyle of the monks was kept secret to outsiders. Guests were allowed to visit and stay only in the guest quarters of the monastery. Strict measures were taken to assure that visitors were kept separated from the majority of the brethren. Workshops and sacred portions of the monastery were off limits unless the guests themselves were monks. The Rule of Saint Benedict forbade monks either to associate with or speak to the visitors.[145] The monastery's perimeter was difficult to cross; only with the permission of the abbot could the brethren breach the monastic enclosure. Those who received permission to leave were forbidden to tell the outside world what went on inside the walls of the monastery and were also not permitted to tell the brethren what went on in the outside world.[146]

The Rule of St. Benedict, including its provisions regarding monastic labor practices and the enforcement of silence and secrecy, influenced the architecture of Benedictine monasteries and cloisters. Indeed, their architecture was based on the Rule of St. Benedict.[147] These monasteries and cloisters manifested the varying degrees of privacy required for different aspects of Benedictine life. Individual monks' cells in the cloisters guaranteed privacy, encouraged meditation, and enforced silence.[148] Visitors were prohibited from entering the cells. Monks were only occasionally permitted to enter another monk's quarters. Artisanal workshops were usually located either immediately outside the compound of the monastery (yet still on the monastery's property) or at its periphery.[149] Seminaries, where communication of thoughts and ideas was paramount, were designed in a classroom-like fashion. The monastery's library was always located near the monks' cells and seminary for easy access. Artisans from the surrounding communities who did not belong to the order could not enter the basilica, the most sacred portion of the monastery or cloister. In short, the Rule of Saint Benedict required a sharp delineation, both temporally and spatially, between the different aspects of the monks' lives. The daily routine of reading, worship, meals, labor, and silence was rigidly plotted out by the Rule.[150] Similarly, there were proper places where the monks ate, meditated, worked, and worshiped. Benediktbeuern, like all Benedictine cloisters, was made up of spaces with varying degrees of public access. This extensive, highly controlled space proved crucial to Fraunhofer's experimental design.

In recent years, a small yet significant literature describing social life and its practices as contingent upon their spatial organization has emerged.[151] There are two ways in which the dichotomies of public/private and openness/restriction can be fruitfully analyzed.[152] The first is to think in terms of social space. Social space assigns appropriate places for practices. It ensures continuity, cohesion, and incorporation of those practices; permits analysis of a particular community or culture; and establishes restricted areas.[153] Where one worked and what one did in Benediktbeuern certainly depended upon who one was. The second way is to consider the notion of boundary. Boundaries separating social spaces are informative insofar as historians can probe the permeability of such borders.[154] Where secrecy is not an issue, such boundaries tend to be negotiable. Boundaries and their junction points are often sites of controversy and of friction. They will, therefore, vary with the particular type of culture the historian is analyzing. Such an analysis is very similar to Steven Shapin's pioneering paper on Robert Boyle's laboratory.[155] In this work, Shapin draws a compelling connection between the empiricist processes of knowledge-making and the spatial distribution of the participants. By looking at who could go into certain social spaces and how the entry into such spaces was regulated, Shapin illustrates how trust is crucial to the evaluation of experimental knowledge. Similarly, in a more recent article about Andreas Libavius's Chemical House, William Newman has convincingly argued how important the social space of a laboratory is in establishing a gradient of privacy.[156] As was the case in Benediktbeuern, certain spaces of Libavius's fictional Chemical House were accessible only by Libavius himself. The architecture of the Chemical House enabled him to differentiate between public and private knowledge.

The notion of social space and its analysis add to our attempt to understand the scientific enterprise as a cultural activity. This account of Fraunhofer's production of achromatic lenses makes it clear that Fraunhofer and Utzschneider were rather skilled at manipulating a culture in which architecture could be put to practical use. Benedictine architecture instantiated a gradient of privacy that was precisely what Utzschneider and Fraunhofer needed in order to guarantee their optical monopoly. Just how German and British experimental natural philosophers reacted to Fraunhofer's private, secretive artisanal practices will be the focus of the ensuing three chapters.

4

The German Response to Fraunhofer's Private Knowledge

The Academy must not become a corporation of artistes (Künstler), factory owners (Fabrikanten), and artisans (Handwerker).
—Joseph von Baader, 1820

Much of the secondary literature has underscored Fraunhofer's success and recognition as a Naturwissenschaftler (experimental natural philosopher). Such accounts often give the reader the impression that, despite his humble origins and his artisanal secrecy, Fraunhofer's work was universally recognized as belonging to a higher class of experimental natural philosophers.[1] Fraunhofer, a working-class artisan, put Bavaria on the scientific map. Although late-nineteenth-century accounts praised the sagacity of German governments that recognized and coordinated the labor of skilled artisans such as Fraunhofer, did Fraunhofer enjoy a privileged status during his lifetime? Was he considered an equal among German investigators of nature and mathematicians during the 1810s and the 1820s, or was he simply considered a skilled artisan? Gauss, Olbers, Schumacher, Bessel, and other astronomers purchased astronomical instruments made in Benediktbeuern, but did they consider the manufacturer of the lenses of those instruments an intellectual equal, or a mere craftsman? Were the skills possessed by Handwerker such as Fraunhofer part and parcel of the scientific enterprise? Should Handwerker receive the same recognition as astronomers, physicists, and mathematicians? This was precisely the issue that was being debated by experimental natural philosophers in several German states and throughout Great Britain during this period. It is certainly clear that well before the 1810s instrument makers were claimed to be crucial to the scientific enterprise throughout both the German states and Great Britain. But whether the knowledge that skilled artisans generated could be reproduced by the rational principles of experimental natural philosophers was a point of contention. Was the knowledge that skilled artisans possessed somehow unique, and was that knowledge subordinate to the knowledge of the principles of science familiar to the experimental natural philosopher?

Before discussing the attitude of several savants to Fraunhofer's work, it is necessary to see how the German Republic of Letters defined the role of authorship, including scientific authorship, around 1800. Late in the eighteenth century, a literary culture was finally blossoming in the German territories, lagging far behind Britain and France. One result of this rapid transformation of the literate middle class was a bifurcation in German literary circles between "highbrow" authors who identified with Friedrich Gottlob Klopstock's *Deutsche Gelehrtenrepublik* and authors who catered to the predilections of "the masses." Many of the "lowbrow" authors needed to generate funds to sustain their humble existence; hence, simplistic tales and self-plagiarism (the process of regenerating themes in order to write more numerous works) were rather common.[2] Authors enjoying less challenging financial circumstances, such as Johann Wolfgang von Goethe (who earned his keep as Privy Councilor to Duke Carl August of Weimar) frowned on those who gave in to the base whims of the public.

That succumbing to the predilections of the lower classes destroys creativity was a powerful sentiment in the German territories around 1800. Artisanal knowledge had historically been considered to be the antithesis of inspired genius. For example, during the Renaissance being a craftsman, or a "master of a body of rules or techniques,"[3] had been deemed to be one of the two necessary components of authorship. Craftsmanship had been defined in contrast to authorship's other quality, inspiration or genius. Inspiration was seen as being creative, a higher form of knowledge not shamefully following the rules or techniques required of the craftsman. As time went on, the role of the author as craftsman began to wane; in the late eighteenth century, theorists exclaimed that it had been totally eclipsed by inspirational genius. Following Edward Young's claim that "imitations are often a sort of manufacture wrought by those machines, art and labor, out of pre-existent materials not their own,"[4] the German literary intelligentsia saw the author as the transcender of rules. Indeed, between 1773 and 1794 a debate over the ownership of intellectual property flared throughout the German territories, sparked by Klopstock's announcement that authors should circumvent publishers and present their work directly to the public via subscription.[5] The critical shift in defining the author as a creative, inspired genius was accompanied by the belief that the author had rights; the message of the book was the author's rather than the audience's. When authors had been seen as being mere craftsmen, the book dealers and publishers had been the owners of the knowledge presented in the text. The debate culminated with a lengthy treatise by Ernst Martin Gräf, titled *Forschungsbericht: Versuch einer einleuchtenden Darstellung des Eigenthums des Schriftstellers und Verlegers und ihrer gegenseitigen Rechte und Verbindlichkeiten. Mit vier Beylagen. Nebst einem kritischen Verzeichnisse aller deutschen besonderen Schriften und in periodischen und anderen Werken stehenden Aufsätze über das Bücherwesen überhaupt und den Büchernachdruck insbesondere* (*A Research Report: An Attempt Toward a Classification of the Property and Property Rights of Writers and Publishers and Their Mutual Rights and Obligations. With Four Appendices. Including a Critical Inventory of All Separate Publications and Essays in Periodicals and Other Works in German Which Concern Matters of the Book as such and Especially Reprinting*). Gräf's *Forschungsbericht* sided with Klopstock's *Deutsche Gelehrtenrepublik*: the authors should be the owners of their work. Gräf wanted

> To ascertain whether it might be possible by arranging such subscriptions for scholars to become the owners of their writings. For at present they are so only in appearance, book dealers are the real proprietors, because scholars must turn their writings over to them if they want to have these writings printed. This occasion will show whether or not one might hope that the public, and the scholars among themselves, . . . will be instrumental in helping scholars achieve actual ownership of their property.[6]

As obvious as Klopstock's plea might sound, before this period, throughout the German territories, a book had been seen as a collaborative enterprise, each group of artisans receiving the same amount of credit as the others. For example, in 1753 the *Allgemeine Oeconomsiches Lexicon* listed all "the artisans" responsible for producing a book in its entry for "Book": "the scholar and writer, the papermaker, the type founder, the typesetter and the printer, the proofreaders, the publisher, the book binder." They had been all equally deserving of credit of the manufacture, authorship, and ownership of the book's contents. Klopstock's intervention attempted to thwart mid-eighteenth-century egalitarianism by granting ownership exclusively to the author, who was no longer considered to be part artisan.

Klopstock's trials were to be rewarded, but not until 1810, after the Napoleonic occupation of the German territories. Baden jurists added laws covering literary property to the Code Civile:

> §577.da. Every written transaction is originally the property of the person who composed it, as long as he did not write it on the commission of another and for the advantage of another, in which it would be the property of the person who commissioned it.[7]

Fraunhofer's Bavaria defined the object of the author's proprietary rights in 1813 by drawing upon works of Klopstock and of Johann Gottlieb Fichte and on Article 397 of the Bavarian Penal Code:

> Anyone who publicizes a work of science or art without the permission of its creator, his heirs, or others who have obtained the rights of the creator by reproducing it in print or in some other way without having reworked it into an original form will be punished.[8]

These debates about authorship spilled over into debates about scientific authorship during the 1820s. In short, artisans were rarely granted the status of experimental natural philosophers for three reasons. First, the importance of secrecy to the artisanal trade was seen as anathema to the Republic of Letters, whose members prided themselves on the openness of scientific knowledge. Second, savants were reluctant to accept artisans as their intellectual equals since craftsmen were members of a commercial nexus; financial interests tainted their work.[9] Third, members of the Republic of Letters argued that artisans manipulated pre-existing materials; they did not create anything. This slavish following of craft rules was deemed the antithesis of creative, scientific knowledge.

But the manual skills associated with mechanics and glassworkers were still considered important enough to be taught at the university level (at least at Jena), and skilled artisans with honorary degrees, rather than professors of physics and chemistry, were the instructors. Jena was not the only university that recognized the importance of artisanal knowledge to the physical sciences. In 1819 Fraunhofer was made an honorary titular professor at the Royal Academy of Sciences in Munich, and in 1822 he received an honorary doctorate from Erlangen.[10]

Although it would be tempting to argue that all Germans realized the importance of artisans and rewarded artisanal labor while the British either exploited or ignored such knowledge, national contexts are too vague and simplistic to be helpful here. Indeed, within certain circles in Fraunhofer's

QVOD FELIX FAVSTVMQVE ESSE IVBEAT
DEVS OPTIMVS MAXIMVS
SVB AVSPICIIS
AVGVSTISSIMI ET POTENTISSIMI REGIS ET DOMINI
DOMINI
MAXIMILIANI IOSEPHI
REGIS BAVARIAE
REGIS ET DOMINI NOSTRI LONGE CLEMENTISSIMI
EX DECRETO AMPLISSIMI PHILOSOPHORVM ORDINIS
IN ACADEMIA REGIA FRIDERICO-ALEXANDRINA ERLANGENSI
PRORECTORE MAGNIFICO
VIRO ILLVSTRISSIMO ET EXPERIENTISSIMO
D. ADOLPHO CHRISTIANO HENRICO HENKE
AVGVSTISSIMO BAVARIAE REGI AB AVLAE CONSILIIS MEDICINAE PROFESSORE PVBLICO ORDINARIO ET INSTITVTI
CLINICI DIRECTORE
VIRO PRAENOBILISSIMO ATQVE DOCTISSIMO
IOSEPHO FRAUENHOFER
INSTRVMENTORVM OPTICORVM FABRICAE QVAE MONACHII FLORET DIRECTORI SOLERTISSIMO
VIRO INGENII SVBTILITATE DOCTR[illegible] ARTE MAXIME CONSPICVO
HONORIS CAVSA
DOCTORIS PHILOSOPHIAE ET AA. LL. MAGISTRI
GRADVM IVRA ET PRIVILEGIA
DIE XVIII. OCTOBRIS [illegible]
RITE CONTVLIT
D. THEOPHILVS ERNESTVS AVGVSTVS MEHMEL
AVGVSTISSIMO BAVARIAE REGI AB AVLAE CONSILIIS PHILOSOPHIAE ET AESTHETICES PROFESSOR PVBLICVS ORDINARIVS
BIBLIOTHECAE ACADEMICAE DIRECTOR ET FACVLTATIS PHILOSOPHICAE H. T. DECANVS
PROMOTOR AD HVNC ACTVM LEGITIME CONSTITVTVS

Figure 4.1
Fraunhofer's honorary doctorate from Erlangen, October 1822. *(Deutsches Museum, Munich)*

Figure 4.2
The Jesuit College in Munich, c. 1620. During the early nineteenth century it was the site of the Royal Academy of Sciences. *(Deutsches Museum, Munich)*

own Bavaria, whether artisanal knowledge counted as scientific knowledge and whether instrument makers should be considered scientific authors were topics of debate, particularly within the confines of the Royal Academy of Sciences in Munich. Fraunhofer's nomination to the rank of ordinary visiting member of the Mathematics and Physics Section of the Academy can be used to trace the contours of the debate over the status of artisanal labor and its relationship to the production of scientific knowledge. The debate centered around the argument of whether Fraunhofer was a Naturforscher (investigator of nature) or a geschickter Handwerker (skilled artisan).

In 1820 Fraunhofer was proposed for a promotion from corresponding member, which he had been since 1817, to ordinary visiting member. The recommendation stated:

> Herr Professor Fraunhofer has become famous among physicists over the past several years for his direction of the Optical Institute formally in Benediktbeuern, now in Munich. His secret of the manufacture of flint and crown glass and the production of optical glass of a size hitherto unheard of have secured an everlasting name for him in the history of science.[11]

The letter continued by praising Fraunhofer's sharp sense of observation, which had led to discoveries in the field of optics.[12]

This recommendation elicited an immediate protest from the Director of Machinery of the Prince's Coin and Mining Office (Maschinendirektor beim kurfürstlichen Münz- und Bergmeisteramt), Joseph von Baader, an ordinary member of the Mathematics and Physics Section. Baader was an engineer and mechanic. After receiving his medical degree, he had turned his attention toward more technical subjects, particularly mining. He was to play a major role in the planning of railroad tracks throughout Bavaria in the 1830s, just before his death. Perhaps best known for the Baader pump (used in mining), he was the author of numerous essays in learned journals dealing with a range of subjects, including suction pumps, hydrolic machines, and train tracks. In his letter to the Academy dated 31 March 1820, Baader complained that Fraunhofer's reputation was insufficient for an ordinary membership. He quoted the Academy's constitution (paragraph XIII, title 1), which stated that ordinary members may be accepted only if the world of scholars has been convinced of the merit of the potential member's published works, or if the Academy has been privy to important discoveries made by the potential member in lectures.[13] Baader continued by emphasizing that Fraunhofer was not university educated and indeed had never attended a Gymnasium. Although Fraunhofer was admittedly well versed in the Kunstfach (art) of practical optics as a result of his training in optical glassmaking, this knowledge was insufficient for Fraunhofer to be called a mathematician or a physicist.[14] Baader's mean-spirited attack became most vitriolic when he warned: "The Academy must not become a corporation of artistes (Künstler), factory owners (Fabrikanten), and artisans (Handwerker)."[15]

But Baader's diatribe did not stop there. He proceeded to attack Fraunhofer's article "On the Determination of the Refractive and Dispersive Indices of Differing Types of Glass" by asserting that, although the article was very interesting and useful for artisans working on the perfection of optical instruments, it lacked any form of a scientific discovery (wissenschaftliche Entdeckung). The dark lines of the solar spectrum were merely an artisanal finding. Baader even questioned whether Fraunhofer had written the article.

Since his essay had been published in the Academy's *Denkschriften*, Fraunhofer had fulfilled one of the necessary (and sufficient) conditions for appointment as a regular member of the Mathematics and Physics Section. In order to disqualify Fraunhofer's candidacy, Baader publicly questioned whether Fraunhofer actually wrote the piece himself, strongly suggesting that the true author was his employer Utzschneider. Clearly, questions of authorship were deeply rooted in issues of social class. Baader concluded his letter by arguing that the secretive nature of glassmaking is "not of a scientific nature, but of an artistic, artisanal one."[16] He argued that Fraunhofer's decision not to disclose his glassmaking techniques disqualified him from the Academy.[17] Fraunhofer's private knowledge was, in Baader's eyes, the antithesis of science.

Baader's protest was joined by an attack on Fraunhofer's character by Julius Konrad Ritter von Yelin, Munich's chief financial advisor (Oberfinanzrat) and a professor of mathematics and physics. His assault was brief, but just as harsh as Baader's. Yelin echoed Baader's concern that Fraunhofer was not university educated. Such a lack of formal education, Yelin argued, would result in an inability to follow the complex lectures that periodically took place in the Mathematics and Physics Section of the Academy. Yelin's anger was most evident in his concluding remark: he found it personally insulting that Fraunhofer would join the same section, at the same rank, as he.

Yelin was not opposed to the application of technology in the service of the state. Quite the contrary, he extolled the progress made by Bavarians in science and technology and how such progress had strengthened Bavaria. Indeed, Yelin played a critical role in Bavaria's Polytechnischen Verein, which had the express goal of applying scientific and technological advances to the fledgling Bavarian economy.[18] What Yelin objected to was that those responsible for such progress should necessarily be considered Naturforscher.

Baader's and Yelin's resistance sparked Fraunhofer's supporters in the Academy to take a concerted action. They argued that there had been historical precedents whereby men without a university education had become ordinary members of the Academy.[19] Indeed, the section's secretary and botanist, Franz von Paula Schrank, composed a memorandum listing members of other academies who did not possess a university education. Johann Georg von Soldner defended his friend's work.

Soldner was the scientific director of the Bavarian ordnance surveying project and the organizer of the Royal Observatory in Bogenhausen, near Munich. His mathematical abilities, evidenced by his *Theorie der Landes-*

vermessung, were well known. He often aided Fraunhofer when the young optician sought his mathematical acumen. He was named an ordinary member of the Royal Academy of Sciences in Munich in 1813. In 1815 he was appointed Court Astronomer, where he contributed to the theory and practice of astronomy and geodesy. In 1818 he tested Fraunhofer's craftsmanship by ordering Fraunhofer's achromatic telescopes for his observatory. Of all the members of the Academy, Soldner was the most qualified to speak on Fraunhofer's behalf.

Soldner simply could not agree with Baader's claim that Fraunhofer's work did not contain a single scientific discovery. The dark lines of the solar spectrum, for Soldner, were such a discovery.[20] At the actual vote, he continued his plea: "Through these lines *exact measurements* of the solar spectrum are now possible, and the possibility of exact measurements and their implementation is the goal of what one considers to be *exact science*. I consider this discovery of Fraunhofer's to be the most important one in the area of light and colors since Newton."[21] Hence, the Academy should not bar Fraunhofer based on the grounds that his work did not belong to a recognized scientific canon. By placing Fraunhofer's name in the same sentence as Newton's, Soldner was indeed considering Fraunhofer to be a great physicist and mathematician.

In the end, it was decided that appointing Fraunhofer to an ordinary visiting membership was too controversial. With only one vote cast against him, he was promoted from a corresponding member to an extraordinary visiting member of the Academy.[22] As an extraordinary visiting member, he was at least permitted to attend sectional meetings. In that same year, Fraunhofer's most theoretical piece hitherto was accepted for publication in the Academy's journal, *Denkschriften der königlichen Bayerischen Akademie der Wissenschaften*. It was titled Neue Modification des Lichtes durch gegenseitige Einwirkung und Beugung der Strahlen (New Modification of Light through Reciprocal Effects and Diffraction of the Rays). Soldner took the lead in proposing the paper to the *Denkschriften*. He argued that it marked "a new epoch in the physical theory of light."[23] Other Academy members agreed, including Reichenbach; the physicist, Benedictine monk, and later professor at the University of Munich Thaddäus Siber; Schrank; and even Baader. Yelin protested again, arguing that the work of someone with so little formal education should not be included in such a prestigious journal.[24] This time, his protest was to no

Königliche Akademie
der Wissenschaften in München

In Folge eines Beschlusses der K. Akad. der Wissenschaften, nach den bestehenden Wahlformen, und nach
erhaltr. Genehmigung Sr. Königl. Maj. von Baiern ist

Herr Joseph

Fraunhofer

zum ausserord. besuch. Mitgliede der Akademie ernannt,
und dessen Name in ihre Listen eingetragen worden.

München den 27ten Jun. 1821.

Schlichtegroll Moll.

Figure 4.3
Fraunhofer's extraordinary visiting membership to the Royal Academy of Sciences in Munich, June 1821. *(Deutsches Museum, Munich)*

avail. In 1823 Fraunhofer was appointed Konservator and professor of the Academy's collection of mathematical and physical instruments, with a stipend compensating him for lectures periodically delivered at the Academy until his death in 1826.[25] He was also elected member of both the Gesellschaft für Naturwissenschaften und Heilkunde of Heidelberg and the Kaiserliche Leopoldinische Akademie. In addition, he was knighted by King Maximilian I in 1824.[26]

It is clear, then, that Fraunhofer's status was hotly contested during the third decade of the nineteenth century in Bavaria. Generally, it was the reform-minded officials of King Maximilian I's government, such as Montgelas and Utzschneider, who appreciated the importance of the labor of artisans during this period, after decades of neglect. As the Swiss historian Heinrich Zschokke claimed in 1817, previous to Maximilian I's rule, Bavaria was the most underdeveloped region among the German states.[27] Zschokke credited Maximilian for restoring to Bavaria the cultural and scientific status it had attained during the Middle Ages. In particular, he congratulated the king's decision to restore the wealth of Bavaria by recreating a community of artisans (Handwerker) that had fled to Italy and the Tirol after Bavaria's devastation during the Thirty Years War.[28] The reforms of Montgelas transformed Bavaria into a powerful manufacturing and trading nation. Such reforms, Zschokke argued, brought scientific and economic prestige to a people who already enjoyed a rich cultural heritage. Indeed, he argued that of all the new business enterprises springing up throughout Bavaria the most noteworthy was Utzschneider and Fraunhofer's Optical Institute.

As we have seen, Fraunhofer's private knowledge, in the form of his skilled manipulations of and recipes for optical glass manufacture, was a double-edged sword. In one obvious sense, it was very beneficial to Fraunhofer, as it guaranteed his Optical Institute's monopoly, thereby increasing his annual salary. However, in another sense, his private knowledge was rather deleterious to his efforts to gain the recognition enjoyed by Naturwissenschaftler. The secrecy shrouding his glass manufacture stood in sharp contrast to the opinions of several Bavarian Naturwissenschaftler on what science ought to be. There were some individuals during the early nineteenth century who still wished to differentiate between scientific and artisanal knowledge. Yelin and Baader argued that his artisanal knowledge was not sufficiently creative enough to be considered scientific knowledge. They

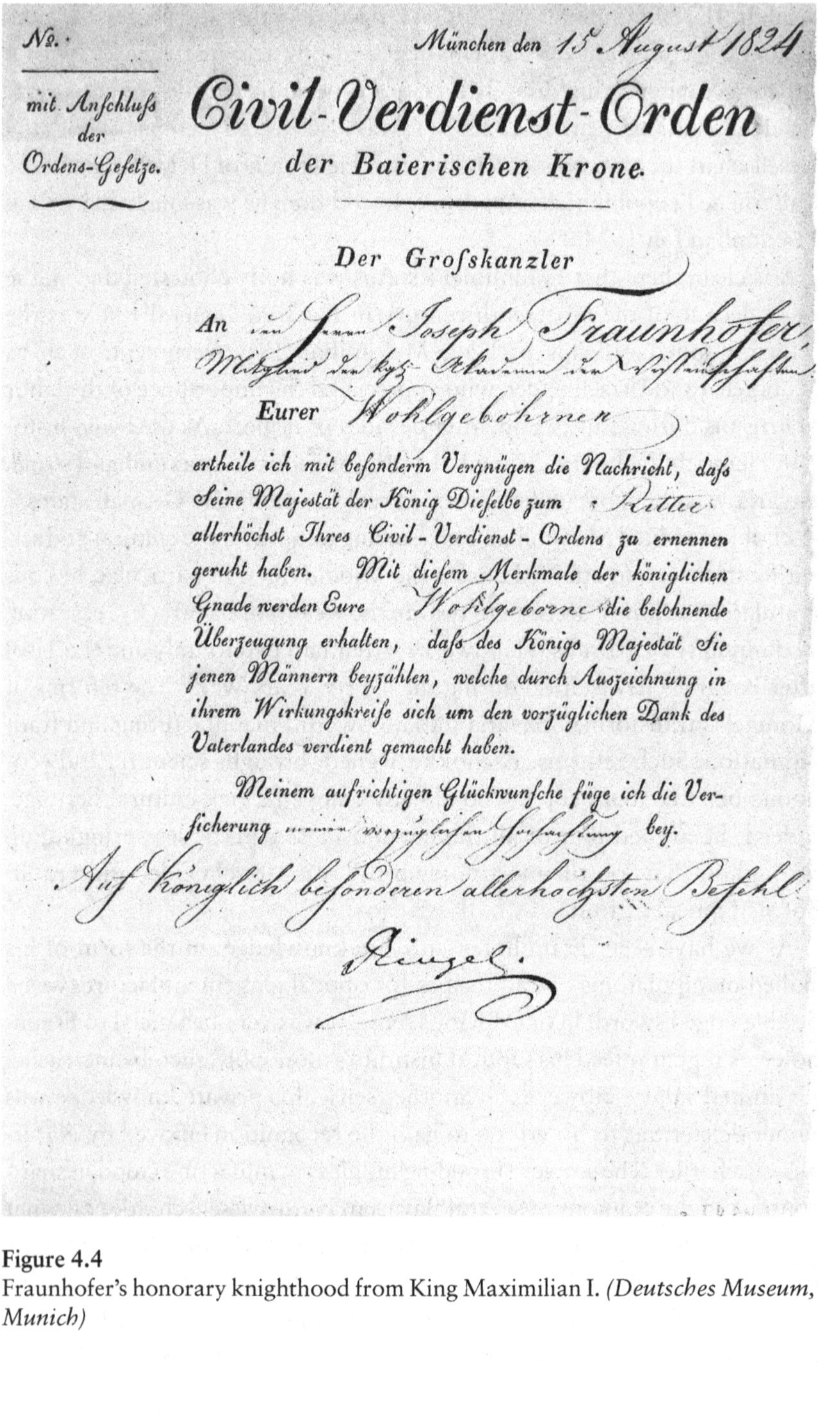

No. München den 15 August 1824

mit Anſchluſs der Ordens-Geſetze.

Civil-Verdienst-Orden
der Baierischen Krone.

Der Groſskanzler

An Herrn Joseph Fraunhofer, Mitglied der kgl. Akademie der Wissenschaften

Eurer Wohlgebohrnen

ertheile ich mit beſonderm Vergnügen die Nachricht, daſs Seine Majestät der König Dieſelbe zum Ritter allerhöchst Ihres Civil-Verdienst-Ordens zu ernennen geruht haben. Mit dieſem Merkmale der königlichen Gnade werden Eure Wohlgeborne die belohnende Überzeugung erhalten, daſs des Königs Majestät Sie jenen Männern beyzählen, welche durch Auszeichnung in ihrem Wirkungskreiſe sich um den vorzüglichen Dank des Vaterlandes verdient gemacht haben.

Meinem aufrichtigen Glückwunſche füge ich die Verſicherung meiner vorzüglichen Hochachtung bey.

Auf königlich beſonderen allerhöchsten Befehl

Figure 4.4
Fraunhofer's honorary knighthood from King Maximilian I. *(Deutsches Museum, Munich)*

did not want to include Fraunhofer among the ranks of Naturwissenschaftler, because he was not university educated and he never disclosed his methods of achromatic-lens production. Secrecy was a characteristic of the market, not the academy. Fraunhofer's secrecy was also a point of contention in Britain during the 1820s and the 1830s. That Fraunhofer would not communicate his knowledge of optical glassmaking, if such knowledge was indeed communicable, deeply concerned leading British experimental natural philosophers as they attempted, in vain, to recapture the optical market.

Thus, for Bavarian bureaucrats, and numerous Naturwissenschaftler, Fraunhofer was a fellow Naturwissenschaftler because of his superior glass, which enabled him to reveal the solar lines of the spectrum. The discovery of the dark lines garnered him the privilege and honor associated with membership to the Academy. Indeed, his knighthood, honorary doctorate, honorary professorships, and memberships in various academies make for a rather impressive Lebenslauf. However, there were those who attempted to thwart Fraunhofer's relentless search for scientific credibility. The disdain of a portion of the Gelehrtentum for the working class was rather apparent, since skilled artisans were by and large not university educated, and their knowledge was necessarily secretive. Just who could belong to this so-called open Republic of Letters was at stake, and Fraunhofer fared far better than most.

5

The British Crisis

Dr. [William Hyde] Wollaston gave me a little prism, which is doubly valuable, being of glass manufactured at Munich by Fraunhofer, whose table of dark lines has now become the standard of comparison in that marvelous science.

—Mary Somerville, circa 1820

There is no country where labor is so divided as here [in Britain]. No worker can explain to you the chain of operations, being perpetually occupied with one small part: listen to him on anything outside that and you will be burdened with error. This division is well intentioned, thus resulting in inexpensive handwork, the perfection of the work and the security of the property of the manufacturer.

—Le Turc, in letter dated 4 September 1786[1]

Throughout the eighteenth century Britain had been the undisputed leader in glass manufacturing and optical technology, but by the second decade of the nineteenth century Britain's lead had been usurped by Fraunhofer and the Optical Institute of Bavaria. As Bavaria was establishing an optical glassmaking community, the British excise tax on crown and flint glass and the labor aristocracy associated with the Industrial Revolution were impeding attempts by the British to improve the quality of their optical lenses.

The first section of this chapter discusses the decline of the British glass industry occasioned by the excise duty, which also crippled optical lens research during the last half of the eighteenth century and the first half of the nineteenth. Since these excise laws required that taxable glass (such as crown and flint glass) be manufactured separately from duty-free glass (such as bottle glass), the skills and practices associated with optical glassmaking became less and less common. Indeed, a labor aristocracy was created which resulted in specialization within the glass industry. This separation of skills rendered futile attempts by Michael Faraday, John

Herschel, and George Dollond to reconstruct the labor practices associated with Fraunhofer's optical lenses.

During the 1820s, British experimental natural philosophers became painfully aware of how inferior their glass and their lenses were in comparison with Fraunhofer's. Indeed, the superiority of Fraunhofer's achromatic lenses formed yet another chapter in the woeful saga of the decline of science in England. The British response, analyzed in the remaining sections of this chapter, can be divided into three attempts to recreate Fraunhoferian lenses, each attempt more frantic than the previous one. The well-orchestrated first attempt involved the establishment of the Joint Committee of the Board of Longitude and Royal Society for the Improvement of Glass for Optical Purposes in April of 1824. This committee and its subcommittee included some of Britain's leading experimental natural philosophers. It was the explicit goal of the subcommittee to procure lenses that could match or better Fraunhofer's. The second attempt to regain world hegemony in optical lenses was the construction of liquid lenses, a technology invented by Robert Blair in late-eighteenth-century Scotland which used a sulfuret of carbon (now called carbon sulfide), instead of a flint-glass lens, to correct chromatic aberration. Peter Barlow, with large subsidies from the Joint Committee, resurrected this technology in 1827 as an attempt to compensate for the inferiority of British flint glass. The third attempt, which was considerably less honorable than the previous two, resorted to bribery in order to obtain Fraunhofer's and Guinand's secrets of producing achromatic lenses. The British offered lucrative pecuniary rewards to the relatives of Pierre Louis Guinand, to Fraunhofer himself, and to French opticians who had obtained Guinand's secrets from his son, Henri. Unfortunately for the British, all three attempts failed. They would not regain their optical monopoly until the 1850s.

Britain's Glass Industry

Bavaria's rise to optical fame was concurrent with a drastic deterioration in the British glass industry, with the decline of the status of the skilled artisan, and with the existence of a labor hierarchy that impeded the exchange of information between ordinary glassmakers and opticians. Britain, which had supplied the world with scientific instruments, now turned to Bavaria for ways of improving its own optical technology. Since a proclamation by James I in 1615, wood had been banned for use as fuel in glassmaking. Glass factories had been rapidly depleting the nation's timber supply, thereby threatening the strength of England's navy, which, of course, needed the wood for shipbuilding.[2] Coal was to become the major fuel of glasshouses. In the year of King James's decree, Sir Robert Mansell had been granted a patent for the use of coal as a fuel in glassmaking.[3] As a result, seventeenth-century England had witnessed a migration of glass factories from the counties of Sussex, Surrey, Wiltshire, and Nottinghamshire to major industrial cities such as London, Bristol, Stourbridge, and Newcastle upon Tyne.[4] Those cities enjoyed easy access to coal deposits. London received its coal shipments via the Thames. During the seventeenth century, Bristol had the second-largest port in England, after London. Newcastle benefited greatly from the proclamation, as it was renowned for its coal deposits. Stourbridge also had access to coal and was the site of world-class clay deposits that were used to produce the pots in which the glass was fired. From the late eighteenth century until the middle of the nineteenth, Birmingham joined London, Bristol, Stourbridge, and Newcastle as a glassmaking city.[5]

Not coincidentally, London had possessed the world's finest opticians owing, in part, to the use of coal. Previously, when wood was used as fuel, the crucibles containing the molten flint glass where the fusion of ingredients took place had been open. However, when the furnace was heated with coal as the fuel, the crucibles had to be covered to prevent microscopic bits of coal dust from settling into the pot. This rendered constant stirring problematic, thereby making it more difficult to bring the interior of the crucible to a high enough temperature to ensure homogeneity. Glassmakers eventually learned that a small amount of lead oxide assisted in the distribution of heat.[6] As a result, a fine crystal suitable for tableware was created, as well as a new industry for England, as lead oxide was the key ingredient for optical flint glass.

Until the end of the eighteenth century, English opticians monopolized the production of achromatic objective lenses. In 1785, according to the French savant J.-D. Cassini ("Cassini IV"), the best flint glass was furnished by a London glassmaker named Parker. But shortly thereafter, Parker's chief flint glassmaker retired, and no one could produce such a superior grade of glass again.[7] Cassini IV continued his report on optical glass by discussing the scarcity of flint glass:

> Mr [Peter] Dollond, for his fine telescopes, bought an entire batch from which he selected the most pure part or layer which was without fissures or striae and not gelatinous; he returned the rest. Actually I do not know for what reasons, but it is no longer possible to find a single good piece in the largest masses of material. It is this which makes opticians despair, even the English ones, who rigorously blame their glass manufacturers, who have doubtlessly changed either the form of their furnaces, or their handling of the glass, or even the primary materials.[8]

Although this decline in the quality of flint glass signaled the beginning of Britain's decline in optical glass manufacture, English flint was still better than that of French and German manufacturers until the work of Fraunhofer. Indeed, foreign opticians could, by law, be sold only glass with defects—such as bubbles or striae—thereby safeguarding the interests of the English optical industry.[9] As a result of this practice, French savants attempted to produce crown glass with a refractive index greater than that of flint glass. (Recall the failed attempts of Clairaut's opticians in the 1750s and the 1760s, mentioned in chapter 2.) And during the first decade of the nineteenth century, Jean-Baptiste Delambre and Joseph Louis Gay-Lussac reported the construction of a heavy crystal glass made by the opticians Kruines and Lancon. This heavy crown glass had a greater specific density and greater refractive and dispersive powers than English flint glass. French instrument makers and savants were pleased with the quality of Kruines and Lancon's optical glass. Delambre had an achromatic telescope made from this heavy crown glass.[10] Similarly, Dufougerais (of Mont Cenis) produced heavy crystal glass, but he could manufacture only small optical lenses; his large pieces of glass were not homogeneous.[11]

English optical instrument makers and opticians had enjoyed quite a high social and scientific status, particularly those affiliated with the Royal Society.[12] Recall from chapter 2 that John Dollond had received the Copley Medal, the only medal the Royal Society offered at the time, for his "invention" of the achromatic lens.[13] In 1761 he had been elected a fellow. British instrument makers' fame had been recognized on the Continent as well. English workshops were the leading purveyors of astronomical instruments, with France not offering a challenge to England's monopoly until the last decade of the eighteenth century.[14]

But by the early 1820s, Britain had lost its lead in both glass production and optical technology. This decline is evidenced by the optical instruments found in two leading observatories of the German territories during the second decade of the nineteenth century. Gauss, the director of Göttingen's Observatory, had decided that achromatic telescopes, theodolites, and heliometers supplied by Fraunhofer's Optical Institute were far superior to their British counterparts manufactured by George Dollond, the late Jesse Ramsden, and the Troughton and Simms firm of London. Similarly, the Gotha Observatory, under the direction of the astronomer Bernhard von Lindenau, began switching its orders from London instrument manufacturers to the Optical Institute and the Mathematisch-mechanische Institut of Bavaria during the 1810s and the 1820s.[15] The blockade of 1806 isolating Britain from the Continent, the French occupation and subsequent ordnance surveying of Bavaria, and the rise of an optical community centered around Fraunhofer all contributed to Britain's downfall. Interestingly, as Bennett has shown, the early nineteenth century witnessed a steady decline in the social status of instrument makers in Britain.[16] "Manual labor" became a pejorative term tainted by its working-class connotations. The fall in the social status of instrument makers and opticians reflected a much more encompassing debasement of skilled artisans in Britain. As E. P. Thompson has argued, the glassmaking profession followed the division of labor created by the Industrial Revolution.[17] Technical innovation and the abundance of cheap labor threatened the skilled artisan. An alliance between the aristocracy and the bourgeoisie was fused in order to deprive the working class of its political rights.[18] The power of the state was abused in an attempt to destroy the trade unions.[19]

Concurrent with the decline in the status of instrument makers in England came the decline of the glass factories. The number of glasshouses producing flint and crown glass in London diminished from thirteen in 1696 to three in 1833.[20] The major reason for the rapid decline in England in general, and in London in particular, was the harsh tax levied on glassmakers by King George II's Excise Act of 1746, which impeded glass production until its repeal by Queen Victoria's government in 1845. The excise regulations during the reign of William IV (1830–1837) read:

> [The glassmaker is] to mark and number every workhouse, pot chamber, pot hole, lear, warehouse-room and other place so entered and made use of; and not to use any pot for preparing or making glass without first giving notice thereof to the proper officer. . . . Every annealing oven, arch or lear . . . for annealing flint glass [is] to be made of rectangular form with the sides and ends perpendicular and parallel to each other and the bottom thereof level with only one mouth or entrance,

with a sufficient iron grating affixed thereto and proper locks and keys and other necessary fastenings. Twelve hours' notice [is] to be given before beginning to fill or charge any pot for making glass. . . . Six hours' notice [is] to be given in writing of intention to heat any oven, arch or lear into which any glass is intended to be put for annealing. In all glass chargeable by weight, the grating to the annealing arches [is] to be closed and securely locked and sealed by the officers immediately after all the glass or ware has been deposited therein. In weighing glass the turn of the seale [is] to be given in favour of the crown.[21]

The tax had been increased in 1777, and it continued to be increased throughout the late eighteenth century and the early nineteenth.[22] This tax, which had greatly decreased the amount of glass produced in England, had been charged by the weight of the glass produced. According to an 1832 account, the average quantity of all types of glass "annually retained for home use in the three years ending in 1793 was 373,782 cwt., while the average quantity consumed during the three years ending in 1829, amounted to 364,156 cwt., showing an annual decrease in the manufacture of 9626 cwt., notwithstanding the great increase of population and still greater strides in civilisation. . . . The annual average quantity [of glass] made for home use during the three years ending in 1812 was 413,414 cwt., while the average of the three following years ending in 1815 was 264,931 cwt., being upwards of 35 per cent. upon the larger quantity; a circumstance which could not fail, among other evils, to bring distress and misery upon a considerable number of operative manufacturers."[23]

In 1812 the chancellor of the exchequer imposed an additional duty upon glass manufacturers which resulted in the decrease in glass production by more than one-third throughout the kingdom.[24] The account of 1832 clearly stated how shortsighted government policies toward glassmakers were: "Could any facts more forcibly point out the pernicious tendency of heavy duties upon articles of domestic manufacture, or more clearly indicate the course which it were wise to follow in remodelling to as great an extent and as quickly as is practicable, this branch of our financial system?"[25]

The account concluded with a rather strong admonition to the government explaining the ramifications of their policy:

It is principally owing to these restrictions that so much foreign glass is now brought into this country in the face of what may be considered an amply protecting duty. Foreign manufacturers are allowed to make any and every article out of that quality of glass which will most cheaply and advantageously answer the end, while our own artists are forbidden to form certain objects, except with more costly materials, which pay the higher rates of duty. Nor is this restriction only commercially wrong, since it forms a matter of just complaint on the part of the chemists, that they are unable to procure utensils fitted for effecting many of the nicer operations connected with their science, because the due protection of the revenue is thought to require that such utensils shall be formed out of that quality of glass alone which, apart from all considerations of price, is otherwise, from its properties, really unfitted for the purpose. . . . The manufacturer, and the interests of science are, consequently, made to suffer.[26]

The tax on domestic glass decreased domestic production of glass because merchants bought cheaper foreign finished products. Their domestic counterparts were more expensive because they were made out of so highly taxed glass. "The interests of science are made to suffer" because only flint and crown glass were subject to the tax, as those types of glass were used to manufacture luxury items such as crystal glasses, chandeliers, and optical lenses.

The government enforced the excise tax rather brutally. Two to three officers were quartered in every glasshouse. They were responsible for registering the total weight of glass that was melted and for preventing the removal of any unweighed glass from the premises.[27] When the officers suspected wrongdoing, they would often fit large locks on the entrance doors of annealing kilns and lears. Later, this became standard practice. The officers would fit these locks, at the expense of the glass manufacturer, every Friday or Saturday, and the glasshouse would remain locked until Monday morning. They were also in charge of supervising any machine repairs, which were often necessary in glass manufacture. If an officer was absent when a glass pot broke, the entire glasshouse was shut down. More daring manufacturers would convince their learmen to make the repairs without the officer being informed. If they were caught, the plant would incur large excise penalties. The most annoying practice was the officers' surveillance of the melting pots. The addition of the smallest piece of metal into the pots without the officers' knowledge and consent was penalized by a £50 fine.[28] One glassmaker remarked: "It is astonishing how Flint Glass works exist at all under such a concentration of commercial and manufacturing hindrances as are imposed by the Excise regulations."[29]

A glass manufacturer gave the following report to the Commissioners of Excise in 1833:

Our business and premises are placed under the arbitrary control of a class of men to whose will and caprice it is most irksome to have to submit and this under a system of regulations most ungraciously inquisitorial. We cannot enter into parts of our own premises without their permission; we can do no one single act in the conduct of our own business without having previously notified our intention to the officers placed over us. We have in the course of the week's operations to serve some 60 to 70 notices on these, our masters, and this under heavy penalties of from £200 to £500 for every neglect.[30]

After reviewing the glass excise tax and its enforcement, the commissioners came to three conclusions:

(1) That in their opinion evasion of the duty could not be prevented by any addition to the laws or regulations; (2) That the regulations presented a great impediment and in many cases a complete ban in the way of those experimental researches which are necessary for the adequate pursuit of objects connected with some of the most important branches of Science; and (3) That they also operate as a direct hindrance to our successful competition in the glass trade with foreign countries.[31]

It was to be another 12 years before the government repealed the excise tax. This tax required the glass manufacturer to pay for each glasshouse (approximately £20 in 1826) as well as for a license to produce glass. He also needed to pay a fee pro rata for the weight of the melted glass in the pots as well as a fee proportional to the excess in weight of manufactured glass over 40 percent (later 50 percent) of the determined weight of molten glass.[32] In 1793, the glasshouses in Stourbridge had accumulated penalties totaling over £20,000 per annum. Richardson of Wordsley singlehandedly accrued £1,258 in fines in 1838.[33]

As a result of the excise duty regulations, British glass production was divided into five sections: flint, crown, plate, broad, and bottle.[34] Since flint and crown glass were subject to the tax, they were required by law to be manufactured separately from duty-free glass. Since such stringent regulations were enforced by excise officers, glass factories specialized in glass that was either taxed or free from the duty.[35] Interestingly, different regions of England specialized in different types of glass production. Stourbridge and Birmingham (and the Midlands in general) were renowned for their flint glass. Broad glass was manufactured in the North, particularly Newcastle and St. Helens, while crown glass was produced in Stourbridge and bottle glass in Yorkshire.[36] The division of labor was so extreme, as an article in the *Flint Glass Maker's Magazine* claimed, that "the Crown Glass Makers, the German Sheet Blower, and the Bottle Maker are all confined to one article each and all of them are the same thing over again, there is no variety: everyday's work is but a repetition of the former[,] stays always the same[,] no changing of patterns."[37]

There was a hierarchy among the different type of glassmakers, with the flint glass makers at the top, followed by crown glass makers, flat glass makers, and bottle makers. According to an 1858 article in the *Flint Glass Maker's Magazine*,

> The members of our society may count themselves among those who have the honour of contributing daily to the luxuries of the tables of the nobility of the land, including Her Majesty the Queen. Seeing, then, that we labour at a beautiful art, is it not only our duty and privilege to excel with the same—to be ambitious for our own credit and attainments, and to study taste, richness and beauty?[38]

The glass produced for the working class was seen by the flint glass makers as an insult to their trade. They did not wish "their beautiful trade of glass-making [to] be brought as low as nail making."[39] Although the labor aristocracy resulted in the improvement of certain forms of skilled labor, such as crystal glass production, the partitioning of skilled labor resulted in no one's knowing how to coordinate the practices necessary to produce objects that had to be made of two or more types of glass. (For example, both flint and crown glass were used in optical lenses and prisms.) Flint glass makers were not familiar with the skills and practices involved in crown glass production and vice versa.[40] And English opticians knew nothing of either type of optical glass production, since opticians such as George Dollond did not manufacture their own optical glass. This would prove fatal to the subcommittee's (particularly Faraday's) attempts at reverse engineering Fraunhofer's lenses and prisms.

The excise duty not only crippled the glass manufacturers, it also greatly hindered opticians' attempts to perfect achromatic lenses. As Pellatt reminisced of a time about 25 years earlier: "Neither plate Glass nor bottle Glass manufacturers were subject to the surveillance of the pots: this made it exclusively injurious to the flint glass maker, and was almost a prohibition of alteration of flint, or experiments, and consequent improvements."[41] Flint glass was of prime importance to the production of achromatic lenses. Because of the presence of lead oxide, it has the crucial property of refracting light through greater angles than crown glass. In the middle of the eighteenth century, John Dollond had complained that it was simply impossible to experiment with flint glass for the construction of a perfectly achromatic telescope because of the excise.[42] Crown glass, the other glass crucial for the manufacturing of achromatic lenses which Dollond had used in his experiments, also came under the charge of the excise officer at the Spon

Lane Works of Chance Brothers. An excise law required that all crown glass be less than one-ninth of an inch in thickness, which was too thin for Dollond's research.[43] Indeed, that the excise duty hindered optical research was evident from the fact that the government removed the restrictions occasioned by the excise laws and regulations on glass experiments performed by Faraday and Dollond of the Board of Longitude and the Royal Society subcommittee.[44]

The artisans who manufactured optical glass in Britain, like their Bavarian colleagues, belonged to guilds. Secrecy was paramount in such communities; therefore, written instructions on the manufacture of glass rarely existed. Such practices were usually passed on by oral tradition from master to journeyman (often from father to son). During the nineteenth century, members of the Flint Glass Makers' Society protested against the superfluity of apprentices whom large manufacturers bound or sought to bind to their trade.[45] Too many apprentices, the society argued, would render their highly prized labor practices accessible to a large audience, thereby threatening the society's monopoly. In short, the lack of readily accessible information on optical glass production, coupled with the effects of the excise laws before their repeal and the labor aristocracy present in the glass trade, greatly hampered the British attempt to recapture optical hegemony.

Artisanal Knowledge, Experimental Natural Philosophers, and the Politics of Labor

Early-nineteenth-century Britain witnessed simultaneously rapid changes in both society and the notion of skill that are pertinent to my account of the British response to Fraunhofer.[46] As I shall discuss below, John Herschel, David Brewster, and Michael Faraday were all concerned with the skills and practices involved in the formation of a scientific discipline and how those skills and practices were manifested, and by whom, in society. The attitudes of these experimental natural philosophers toward artisanal skills and practices had ramifications not only for the newly emerging science of physics but also for the newly emerging plans of reform for Britain. These skills and practices became a central issue in major debates of the 1820s and the 1830s which form the political context of this chapter: patent laws, artisanal schools, and (to a lesser extent) mechanization. These attitudes were not idiosyncratic but reflected a wide spectrum of belief during this period. The responses of Herschel, Brewster, and Faraday raise several intriguing questions, such as whether artisanal skills can be replicated by a totally different scientific culture, how artisanal knowledge is communicated, and whether or not it is generally possible to recreate an artifact without witnessing the craft labor that produced it.

David Brewster

David Brewster (1781–1868)[47] entered the University of Edinburgh in 1794. Although he completed his courses, he left the university without taking a bachelor's degree—a common practice of the period. He continued his education as a divinity student and received an honorary M.A. in 1800. In 1804 he was licensed to preach in the Church of Scotland. Brewster was an evangelical, advising leaders of the Disruption. He became a member of the Free Church of Scotland in 1813. He was also an outspoken reform Whig, constantly attacking the English Tories and their privileges.

As was discussed in chapter 2, Brewster's first major work, *A Treatise on New Philosophical Instruments* (1813), illustrates his early fascination with astronomy and optical instruments. In this tome, Brewster determined the refractive and dispersive indices of nearly 200 substances in an attempt to improve the construction of achromatic telescopes. After being made aware of Etienne Louis Malus's work on the polarization of light, Brewster turned away from instrument manufacturing and toward optical theory. He concentrated on the partial polarization of light in mica, the law of polarization by refraction, metallic reflection, and optical mineralogy. His research in these fields resulted, as is well known, in his becoming a staunch proponent of the corpuscular theory of light. Eventually he would come into conflict with John Herschel.

After the 1830s, Brewster turned his attention to photography, stereoscopy, and the physiology of vision.[48] He played a leading role in establishing the Edinburgh School of Arts (1821) and the British Association for the Advancement of Science (1831). He was a fellow of the Royal Societies of London and Edinburgh and a foreign associate of the French Institute. His numerous honors included the Copley, Rumford, and Royal Medals of the Royal Society of London and the Keith Prize of the Royal Society of Edinburgh.

Brewster's first mention of Fraunhofer's work appears in a letter to Heinrich Christian Schumacher, astronomer and director of the Royal

Figure 5.1
David Brewster (1781–1868). *(President and Fellows of the Royal Society of London)*

Danish Academy of Copenhagen. In June of 1821, Brewster thanked Schumacher for procuring "one of the best Munich Achromatic Telescopes."[49] In return, Brewster obtained for Schumacher the best reflecting telescope available in Scotland, made by the Aberdeen optical instrument maker, John Ramage. The purpose of their exchange of telescopes was to make comparisons for the geodetic mapping of Holland, Hessen, and Prussia. Also, Schumacher needed to purchase instruments for the new observatory at Altona, near Hamburg, with funds provided by King Frederick VI of Denmark. On 29 October 1821, Brewster informed Schumacher that he had just received a copy of Fraunhofer's prismatic research from Fraunhofer himself.[50] From that point on, Brewster became deeply interested in the works of the Bavarian optical glassmaker.

As was discussed in the chapter 3, Fraunhofer's work was not published in English until 1823–24, when Brewster translated *Bestimmung des Brechungs- und Farbenzerstreuungs-Vermögens verschiedener Glasarten* (1817) in two parts (volumes 9 and 10) in the *Edinburgh Philosophical Journal*. Interestingly, Brewster added the phrase "with an Account of the Lines and Streaks which cross the Spectrum." It was during 1822 and 1823 that Brewster was attempting to solve the problem of chromatic aberration in microscopes by using monochromatic lamps.[51] He had likewise attempted to overcome chromaticity by using monochromatic light, but Fraunhofer now offered a powerful technique of producing achromatic lenses, which did not rely on a monochromatic light source.

It is noteworthy that no translation was published until 6 years after the paper's original publication in German.[52] It was translated into French in *Schumacher's Astronomische Abhandlungen* in 1823 as well. The English publication of Fraunhofer's now-classic paper signaled a massive reaction to his work in British periodicals. Subsequent articles of Fraunhofer were published in the *Philosophical Magazine and Journal*, the *Edinburgh Philosophical Journal*, and the *Edinburgh Journal of Science*. Brewster gave an account of the Fraunhofer lines in his articles on optics in the *Edinburgh Encyclopaedia*, where he reproduced the spectral lines and Fraunhofer's tables of refraction for differing species of glass and liquids.[53] In 1824 the Astronomical Institution of Edinburgh, a collection of individuals interested in advancing the science of astronomy, wished to furnish their observatory with state-of-the-art astronomical instruments. It was to Fraunhofer and Schumacher that the Astronomical Institution turned. On 20 June

1824, Brewster received a small achromatic telescope made by Fraunhofer.[54] Brewster and the Astronomical Institution were so impressed with Fraunhofer's work that in December of 1824 they decided to inquire about the cost of an "Achromatic Telescope (by Frauenhofer [*sic*]) with Micrometer, complete in all its parts and adapted for astronomical observations. The object glass to be at least 6 English Inches in diameter."[55] They also inquired about object glasses greater than 6 inches in diameter, and the optical part of a telescope 6 English feet in length to be fitted or adapted to a mural circle.[56] One month later, the Astronomical Institution decided that Fraunhofer was to provide their astronomical instruments.[57] Shortly thereafter, the Astronomical Institution authorized Brewster to order a complete transit instrument with an objective possessing a 70-line aperture and an 8-foot focus; two achromatic telescopes identical to the one sent by Fraunhofer 9 months earlier with an added stand for giving a vertical reading; a pair of opera glasses 6 inches in length; and an achromatic lens 1 inch in diameter.[58] Fraunhofer, rather than Dollond, Edward Troughton, or Ramage, was the Scottish choice for arming their Edinburgh observatory.

In 1825 Brewster wrote a brief report for an issue of the *Edinburgh Journal of Science* on Fraunhofer's achromatic refractor with an object lens 9⅗ inches in diameter (9 Parisian inches), which the Russian government had ordered for their Dorpat observatory. We are told that "Frauenhofer [*sic*] is said to have succeeded beyond his most sanguine expectations."[59] Also, in that same issue, Brewster provided a description of this telescope. He begins as follows:

> The greatest discovery of a method of making flint glass in large pieces, and perfectly pure and free from striae, which was made by the late M. Guinand, . . . may be considered as forming an era in the history of the achromatic telescope.
>
> By means of this glass, M. Fraunhofer, the director of the Optical Institute or Manufactory at Benedictbauern [*sic*], near Munich, has constructed achromatic telescopes far superior to any that have hitherto been made; and we can assure our readers, of that which many of them will deem incredible, that this eminent artist can now make achromatic object glasses with an aperture of *eighteen inches*. But it is not merely in the optical part of the instrument that M. Fraunhofer has been successful. His various improvements on the apparatus which accompanies the telescope, and his ingenious micrometers for measuring angles of all kinds in the heavens, have received the sanction of some of the most eminent practical astronomers in Europe, and are now considered as constituting an instrument of incalculable value for general astronomical observations.[60]

In the next year Brewster published the description of Fraunhofer's masterpiece by Friedrich Georg Wilhelm Struve, the astronomer at the

University of Dorpat Observatory. Brewster then attached his comments to the report's conclusion:

Such is the description of Fraunhofer's telescope given by Professor Struve; and we think that no Englishman can read it without feelings of the most poignant regret, that *England has now lost her supremacy in the manufacture of achromatic telescopes,* and the government one of the sources of its revenue. In a few years she will also lose *her superiority in the manufacture of the great divided instruments for fixed observatories.* When these sources of occupation for the scientific talent decline, the scientific character of the country must fall along with them, and the British government will deplore, when it is too late, her total inattention to the scientific establishments of the empire. When a great nation ceases to triumph in her arts, it is no unreasonable apprehension, that she may cease also to triumph by her arms.[61]

In his eulogy of Guinand, Brewster argued that the British excise laws prohibited the manufacture of flint glass in Britain:

The discovery of a method of making Flint Glass for achromatic telescopes has been, during the last seventy years, an object of almost national ambition. In England, unfortunately, the strictness of our Excise laws prevented any attempt from being made on a proper scale to solve this great practical problem; but in France, where no such restrictions existed, numerous attempts have been made to perfect the manufacture of flint-glass for optical purposes.[62]

In his eulogy of Fraunhofer, published in 1827, Brewster translated several sections of Utzschneider's 1826 biography. Once again, Brewster used the conclusion of the article to launch a diatribe against His Majesty's Government. He argued that Bavaria rewarded its instrument makers while the British ignored theirs:

His [Fraunhofer's] own sovereign Maximilian Joseph was his earliest and his latest patron, and by the liberality with which he conferred civil honours and pecuniary rewards on Joseph Fraunhofer, he has immortalized his own name, and added a new lustre to the Bavarian crown. In thus noticing the honours which a grateful sovereign had conferred on the distinguished improver of the achromatic telescope, it is impossible to subdue the mortifying recollection, that no wreath of British gratitude has yet adorned the *inventor* of that noble instrument. England may well blush when she hears the name of [John] Dollond pronounced without any appendage of honour, and without any association of gratitude. . . . The pre-eminence which England had so long enjoyed in the manufacture of the achromatic telescope [is now] transferred to a foreign country. . . . The British minister who shall first establish a system of effectual patronage for our arts and sciences, and who shall deliver them from the fatal incubus of our own patent laws, will be regarded as the Colbert of his age and will secure to himself a more glorious renown than he could ever obtain from the highest achievements in legislation or in politics.[63]

Fraunhofer's death was a grave concern for the Astronomical Institution. Brewster wrote to Schumacher to "solicit any information which [he] possessed respecting the ability of Mr Utzschneider to execute the Transit Instrument which [the Astronomical Institution] ordered on the faith of its being finished under the eye of the late M. Frauenhofer [*sic*]."[64] In short, Brewster wanted to have Schumacher's assurances that the astronomical instruments would still be made with Fraunhoferian quality under the direction of Utzschneider.[65] Although Schumacher apparently did recommend that the Astronomical Institution should allow Utzschneider to complete their order, Utzschneider did not fulfill Schumacher's expectations.[66] On 16 December 1827 Brewster wrote that the members of the Astronomical Institution "are very much annoyed at M. Utschneider's [*sic*] delay in making the lenses, and I begin to fear that he may be endeavouring to give them to some other body."[67] Henry Fox Talbot echoed Brewster's concern. He worried that Fraunhofer's personal knowledge of achromatic-lens production would be lost forever. In a letter addressed to Herschel reporting Fraunhofer's death, Talbot writes: "His method of making Flint glass will be lost to the world as I am told it depended chiefly on his personal attention to the process as it went on."[68] Indeed, Talbot canceled an order for optical lenses from Benediktbeuern, as his confidence in the Optical Institute waned considerably after Utzschneider shipped him an incorrect order.[69]

Fraunhofer's success and the British failure gave Brewster a perfect resource with which to attack the English government, particularly its Tory establishment, and the Royal Society of London. Brewster, after all, was to label the Royal Society "an incubus pressing with its livid weight" on science in consequence of the wave optics controversy.[70] His use of the Fraunhofer episode fits squarely into the "decline of science in England" debate mounted by him and Charles Babbage in the 1820s and the 1830s.[71] Indeed, Babbage mentions Fraunhofer's use of the dark lines in his *Reflexions on the Decline of Science in England* of 1830.

Brewster was concerned by the British government's treatment of mechanical skill and talent. In 1821 he had warned:

> Greatly . . . as do we excel all other nations in the productions of the useful arts, we think it will scarcely be denied that our superiority may still be increased; that a great portion of our manufacturing skill is neutralised by the financial condition of the country; and that the measures of rival industry, which has been rigorously pushed by foreign states, can only be opposed by the most liberal and efficacious excitement of mechanical talent.[72]

That "superiority," Brewster later reported, vanished (by his own admission) by the late 1820s. Brewster criticized the British patent laws for leaving "this country in the singular predicament of being the only nation in Europe which has withdrawn from the safeguard of the law the great products of mechanical skill"[73] and for failing to defend the possessor of mechanical skill, the inventor, from invading pirates.

Mechanical skill, for Brewster, needed to be protected and nurtured in order for Britain to retain its leading position in the mechanical arts. He applauded the Society for the Encouragement of Arts, Manufactures, and Commerce as "the only public establishment which has interposed the labours in behalf of mechanical genius,"[74] adding: "It is deeply to be lamented that Scotland possesses no such institution; and that the profusion of mechanical talent which characterises this part of the island, should be thus allowed to languish in obscurity."[75] Brewster wished to rectify the situation. He offered a portion of his *Edinburgh Philosophical Journal* to inventors so that they could make their views known, and he offered his personal assistance as "a man of science" who "would take the trouble of assisting them with his opinions and advice."[76] Brewster feared that the practical knowledge and the mechanical skill possessed by artisans were being woefully neglected, to the harm of both the craftsman and the general public.

His solution was to establish (with Leonard Horner) the School of the Arts in Edinburgh for the scientific education of Scottish artisans and mechanics, in 1821.[77] The initial aim of the teaching at the Edinburgh School of Arts was "nothing but the application of science to practical arts."[78] Brewster became a member of its board of directors. In that same year Brewster urged the establishment of a Scottish Society of Arts in an attempt to promote and reward Scottish inventors.[79] He was able to enlist numerous Scottish aristocrats to patronize the society. But the School of Arts and Society of Arts competed for students, and Horner was outraged when he learned that Brewster was planning to incorporate the school within his society. Brewster was forced to abandon his Society of Arts before a single course could be offered, and Horner removed him from the board of directors of the School of Arts.[80] During a board meeting where Brewster's intentions were made public, Horner drew an interesting distinction between his school and Brewster's ill-fated society:

> He [Mr. Horner] had never heard that it was any part of the plan of that Society [of the Arts] to give instruction to mechanics, but that it was intended to be on the plan of the Society of the Adelphi in London, for the encouragement of arts and manufactures.[81]

Brewster knew all too well that the practical knowledge of artisans was best taught by master craftsmen in the workplace, and he was interested in the role of such knowledge in manufacturing. But Brewster's dismissal from the board did not detract from his interest in the School of Arts.[82] In 1824 he penned a summary of the school, which belonged to a larger movement of Mechanics' Institutes cropping up throughout Great Britain in the 1820s. According to Brewster, these institutes, devoted to "the diffusion of scientific knowledge among operative mechanics," "ameliorat[ed] the moral condition of the great mass of the people" and "increas[ed] the skill and prosperity of our various branches of manufacture."[83]

Drawing on the work of George Birkbeck at the Andersonian Institution in Glasgow, and later at the London Mechanics' Institute, Horner's school was a place "in which such branches of science as would be useful to mechanics in the exercise of their trade, might be taught at convenient hours, and at an expense that would be within their reach."[84] And in October of 1821, the Edinburgh "School of Arts, for the instruction of Mechanics, in such branches of physical sciences as are of practical application in their several trades" opened its doors, charging its members 15 shillings.[85] Four hundred twenty tickets, each entitling the holder to a year's worth of lectures and admission to the library, were sold. The School of Arts reached its capacity in the first year. Brewster noted:

> The lectures delivered the first year, were on the principles of chemistry and their application to the arts, or the elementary principles of Mechanical Philosophy, or Architecture, and on Farriery. At the close of these lectures at the end of April, one lecture a week on each subject having been delivered during the preceding seven months, there was established a class for Architectural and Mechanical Drawing, which continued for four months.[86]

Brewster's concern that the British government was failing to recognize a chief source of manufacture, the inventors and artisans, also fueled his crusade against the state of the nation's patent laws. As was noted above, Brewster first criticized the patent laws in 1821, in a discussion of the plight of British inventors. He admonished the government that the nation would no longer retain its prominent position in the mechanical arts if the government failed to change the patent laws. And recall that Brewster, in his eulogy of Fraunhofer, claimed that "the British minister . . . who shall deliver us them from the fatal incubus of our patent laws, will be regarded as the Colbert of his age."[87] By 1830 the situation had deteriorated. The views of the radical journalist Thomas Hodgskin, who had argued for the abolition of patent laws in his 1827 book *Popular Political Economy*, were gaining popularity.[88] Hodgskin put forward a determinist intrepretation of invention, claiming that inventions arise due from the needs of a society and that the inventor is merely the right person at the right place who takes the final step in the continual process of ideas—the inventor merely adds the last link to a chain forged by the collective. Brewster upheld the individualist interpretation of invention that underscored the importance and the creative genius of the inventor.[89] In 1830 he penned an essay with his friend and colleague Charles Babbage[90]; his portion dealt with the impact of British patent laws on science. In this piece Brewster contrasts the labor of the "inventor of new machines and the discoverer of new arts" with that of an author:

> He who has invented a new steam engine cannot, like the author of a new romance, dispose of it forthwith. He must devote himself night and day to the practical application of his principle; he must construct models and perform experiments, and he must work either in the dark or with the assistance of tried friends, lest some pirate rob him of his idea, and bring it earlier into use.[91]

Inventing artisans were forced to keep secrets, as the British government could not fulfil its duty in protecting their work:

> A real difficulty exists in protecting the rights of inventors. When a patentee applies for an exclusive privilege, there are two parties whose interests are supposed to be at stake—the *inventor*, and the *public*, as represented by the government. These parties meet on the understood principle, that one has a secret to communicate, and the other a privilege to confer in return. The conditions required from the patentee are, that he really possesses a secret, and honestly communicates it . . . in a document called the *Specification*.[92]

Either the failure of the government to grant the patent or its premature repeal was, in Brewster's opinion, "an act of oppression and dishonesty."[93] It was oppressive because the law in essence deprived the inventor his invention, and dishonest because the government would keep the application fee for the patent (often hundreds of pounds).

Some 20 years later, reflecting back on the 1820s and the 1830s, Brewster stressed that the inventor should keep his work secret, precisely because the patent laws could not protect him if he could not "describe his invention with legal precision":

> If he errs in the slightest point—if his description is not sufficiently intelligible—if the smallest portion of his invention has been used before—or if he has incautiously allowed his secret to be made known to two individuals, or even to one—his patent

will be invaded by remorseless pirates, who are ever on the watch for insecure inventions.[94]

In 1855, Brewster once again emphasized the insecurity of patents:

It almost always happens, that a patentee cannot complete his invention without the advice of friends, and the aid of workman, and many instances have occurred in which his secret has been betrayed, and his rights invaded. He is, therefore, compelled to take out his patent before he can bring it into the market. The attempts of pirates to rob him of his invention, the difficulty of obtaining the aid of capitalists for an untried invention, and many other causes, prevent him from realizing even the amount of his fees till his patent is nearly expired.[95]

Brewster then lambasted the government's claim that patenting a mechanical device was difficult:

It will no doubt be said, that the uncertainty of a patent right arises, in a great measure, from the difficulty of protecting a mechanical invention, or a process in the arts, but this difficulty, or rather this incapacity of our legislators to devise a sufficient protection for the productions of skill, instead of authorizing them to levy a tax upon inventions which they are unable to defend, should have led them to confer bounties or rewards upon those who risk their time and their fortunes in labours which are thus withdrawn from the protection of the law.[96]

Secrecy, then, was the necessary and legitimate defense of honest craftsmen and inventors against ruthless pirates and incompetent legislators. The patent laws harmed not only the craftsmen or the possessor of mechanical genius, but the general public as well:

Many inventions and discoveries are not brought forward by their authors, and many processes in the arts kept secret, on account of the expense, the insecurity, and the short endurance of patent rights. Such processes, too, are often lost to society . . . The public, therefore, suffers a grave injury from the continuance of those causes which induce inventors to withhold their inventions, or the authors of new processes in the arts to work them in secret.[97]

Brewster cited historical examples of how patent legislation hindered scientific progress and cost inventors large sums of money. When Arkwright's patent for spinning machinery was repealed, manufacturers no longer had incentive to improve upon his work, or in Brewster's own words, "the powerful stimulus to invention was removed."[98] And in 1767 Hargrave's patent for the spinning jenny had been invaded by privates.[99]

Following Brewster's cue, Lord Brougham, a prominent Whig politician, championed the cause of inventors by passing minimal reforms through the House of Lords and the House of Commons in 1835. Brewster repeated his concerns over the British patent laws in his Presidential Address to the

British Association for the Advancement of Science in 1850. In 1852 the Patent Amendment Act of Lord Brougham and Lord Granville, the latter of whom supported the determinist interpretation of invention, was passed. But Brewster wanted more.[100] In 1856 Brewster, the Duke of Argyle, and the Earl of Harrowby, on behalf of the BAAS, made three recommendations for changing the patent laws: that patents be granted without any fee, that every patent right be absolutely secured once the patent has been sealed, and that the exclusive privilege for a patent be extended from 14 years to at least 25 years.[101]

Brewster was speaking from his rather bitter and unfortunate experience with British patent laws. Having stumbled upon the principle of the kaleidoscope while conducting experiments on the polarization of light in 1814 and 1815,[102] he had applied for a patent in 1817. It had been an arduous and expensive task, costing him hundreds of pounds. He wrote:

> After the patent was signed, and the instrument in a state of forwardness, the gentleman who was employed to manufacture them under the patent, carried a kaleidoscope to show principal London opticians, for the purpose of taking orders for them. These gentlemen naturally made one for their own use, and for the amusement of their friends; and the character of the instrument being thus made public, the tinmen and glaziers began to manufacture the detached parts of it, in order to evade the patent; while others manufactured and sold the instrument complete, without being aware that the exclusive property of it had been secured by a patent.[103]

Brewster never saw a "farthing" from his patent.[104]

Although Brewster emphasized the importance of mechanical skill to manufacture, he was not concerned with the artisans themselves. In the quote above, he blames "the tinmen and glaziers" for pirating his patent. In an 1831 letter to J.D. Forbes, he seems rather weary of some craftsmen:

> Among the Opticians in London I would advise you to call on Mr R. B. Bate in the Poultry, one of the ablest and certainly one of the best men among them. He has a conscience as well as a head which cannot be said of them all.[105]

In addition to his support of artisanal schools and reformed patent laws for inventors, Brewster applauded treatises, written by experimental natural philosophers, that were intended for the use of skilled laborers interested in practical mechanics. In his preface to the third edition of James Ferguson's *Lectures on Select Subjects* (1823), Brewster congratulated Ferguson for being "the first elementary writer on natural philosophy."[106] "To his labours" he continued, "we must attribute that general diffusion of scientific knowledge among the practical mechanics of this country, which has in a greater measure banished those antiquated prejudices, and erroneous maxims of construction, that perpetually mislead the unlettered artist."[107] For Brewster, such education was a way to transform society by improving both the character of the artisan and the might of British manufacture.

Brewster never believed that the British attempt to produce Fraunhoferian glass would succeed: "We ourselves predicted sixteen years ago, that the committee [of the Royal Society and Board of Longitude] neither would nor could accomplish the object for which they were associated, and we can now record the melancholy truth, that the experimental glasshouse has been long closed, and the experimenters have disappeared."[108] Once again Brewster attacked the British government, his real enemy, for taxing the glass industry. He continued to emphasize the practical knowledge of Guinand in creating achromatic lenses:

But though we have thus lost the monopoly of the achromatic telescope, and are now obliged to import the instrument from rival states, there is nevertheless a law of progress in practical science, with which neither ignorant governments, nor slumbering institutions, nor individual torpor can interfere. What a conclave of English legislators and philosophers attempted in vain, was accomplished by a humble peasant [Guinand] in the gorges of the Jura, where no patron encouraged, and no exciseman disturbed him.[109]

What particularly struck Brewster about Guinand's work was that Guinand was "an old man of seventy-six years of age, who himself manufactured the flint and crown glass, after having made with his own hands his vitrifying furnace and his crucibles, who, without any mathematical knowledge, devised a graphic method of ascertaining the proportion of the curves that must be given to the lenses, afterwards wrought and polished them by means peculiar to himself."[110] Guinand's personal, practical knowledge intrigued Brewster, as the glassmaker found a way to produce lenses without relying on the rational principles of mathematics so cherished by Herschel. In 1832, after reading Faraday's paper "On the Manufacture of Glass" in the *Philosophical Transactions*, Brewster wrote: "I presume you [Faraday] have met with some little difficulties of a practical kind in the formation of large lenses."[111] Brewster continued by asking if Pellatt, other glassmakers, "or ordinary workmen" following Faraday's lead have made the flint glass necessary for Brewster's own optical investigations, as he was interested in patenting a process for new lenses requiring a highly refractive and dispersive glass.[112] Brewster argued that Fraunhofer "united the highest scientific attainments with great mechanical and practical knowledge."[113] This practical knowledge, according to Brewster, was the highly skilled manipulations of artisans that could not be recovered by the theoretical knowledge of experimental natural philosophers. According to Brewster, Fraunhofer improved upon Guinand's work by possessing "a thorough knowledge of chemistry and physics . . . and [inventing] the theory of manipulation, of which Guinand knew only the results."[114] The theory of manipulation—in Brewster's eyes the artisanal component necessary to produce a superior grade of optical glass—was learned by emulation, the apprentice learning it by precept from the master. Brewster was skeptical of the English attempt to produce flint glass of Fraunhoferian quality precisely because he believed that such a feat could never be achieved without recreating Fraunhofer's skillful manipulations, which could not be reproduced by means of theoretical knowledge. For Brewster, the problem the British had to face was producing high-quality flint glass more than 4 inches in diameter.

Practical knowledge, its communicability, artisanal schools, and patent reform were all interrelated themes for Brewster. He reckoned that there was not an explicit philosophical procedure for making glass, but rather a skilled, manual procedure discovered by Guinand and perfected and united with theory by Fraunhofer. His solution to the British glass crisis was to appeal to the canniness of Scottish artisans, particularly Robert Blair.

While Faraday and the subcommittee busied themselves with finding alternatives to flint glass by producing silicated borate of lead, other British natural philosophers were inventing other technologies. Brewster reported an alternative in the *Edinburgh Journal of Science* in 1826. To counter the foreign threat, Brewster turned to the research of his deceased compatriot Blair: "It has often struck us with surprise, that no attempt has been made to introduce and to improve the *achromatic telescopes with fluid object-glasses*, invented and actually constructed by our countryman Dr Blair." Blair had built a telescope "*which equalled in all respects*, if it did not surpass, the best of Dollond's *forty-two inches* long."[115] At a meeting of the Royal Society of Edinburgh held on 7 December 1789, the minutes describing one of Blair's telescopes which Blair had been exhibiting stated:

> Dr Robert Blair, Professor of Astronomy in the University of Edinburgh, produced before the society an achromatic telescope of new construction, in which the convex lenses are of common [i.e., not flint] glass and the errors which arise from the different refrangibility of light, and from the spherical figures of the lenses, are corrected by the interposition of transparent fluids.[116]

As was discussed in chapter 2, in 1791 Blair had publicly communicated his invention to the Royal Society of Edinburgh.[117] By applying several prisms with different angles to a standard prism that contained the fluids to be investigated, Blair had determined the relative dispersing qualities of those fluids. He had successfully discovered a fluid, in which the particles of muriatic acid (now called hydrochloric acid) and metallic particles maintained a certain proportion, that could eliminate the secondary spectrum better than crown and flint doublets could.[118] John Robison of the University of Edinburgh reportedly claimed that Blair had constructed a telescope that did not exceed 15 inches but was equal, if not superior, to George Dollond's 42-inch telescopes.[119] Brewster hoped that liquid lenses might be able to free Britain of dependency on flint and crown glass.

Brewster concluded his 1826 article, not surprisingly, by showing the English the error of their ways: "We earnestly hope that the Royal Society of London, or the Board of Longitude, will take some measures for reviving their manufacture."[120] Brewster's emphasis on the locality of skilled labor, his skeptical attitude toward the role of the philosopher in attempting to emulate the practices and products of the skilled artisan, his support of specialized schools for artisans, and his attempts to reform British patent laws were much in line with Scottish Whig, anti-establishment, anti-Tory, anti–Royal Society, and anti-England politics. His attitude stood in sharp contrast to Herschel's sentiment of universally communicable knowledge.

John Herschel

John Frederick William Herschel (1792–1871; figure 5.2) was the only child of the renowned astronomer, Sir William.[121] He was raised in an upper-middle-class household in Slough where he often ground and polished optical lenses for use by his father. After a brief spell at Eton, Herschel attended a private school nearer his home. At age 17 he matriculated at St. John's College, Cambridge, studying mathematics. While at Cambridge he befriended Charles Babbage, George Peacock, and William Whewell, all of whom would greatly influence British natural philosophy later in that century. They co-founded the Analytical Society, which attempted to bring the advanced methods of mathematical analysis practiced on the Continent to the British curriculum. Herschel sat the tripos examination in 1813 and was

Figure 5.2
John F. W. Herschel (1792–1871). *(President and Fellows of the Royal Society of London)*

named Senior Wrangler. Immediately thereafter he was elected to a fellowship at both St. John's College and the Royal Society. In 1816 he left Cambridge to embark on a scientific career.

After working in mathematics, Herschel turned his attention to physical and geometrical optics,[122] studying the polarization of crystals, the interference of light and sound waves, and the solar system. His first contribution to astronomy (1822) dealt with the computation of lunar occultations. He and James South catalogued 380 double stars in 1824, for which they received the Lalande Prize of the French Academy in 1825 and the gold medal of the Astronomical Society in 1826. After South left England,

Herschel continued his astronomical observations at Slough, improving upon his father's star catalogues. In November of 1833 he and his family sailed from Portsmouth to Cape Town in order to undertake various experiments in astronomy, oceanography, and meteorology. He was an active member of the Royal Society, receiving numerous Copley and Royal Medals and serving as its secretary from 1824 to 1827. He was also a member of the British Association for the Advancement of Science, and its president in 1845, and was one of the founding members of the Astronomical Society (Royal after 1831) in 1820, serving as its secretary from 1820 to 1827 and 1846 to 1847. He was the Royal Astronomical Society's president from 1827 to 1829, 1839 to 1841, and 1847 to 1849, and the gold medalist in 1826 and 1836. In short, Herschel was by the mid 1820s Britain's premier experimental natural philosopher.

Herschel became familiar with the standard of optical glass produced in Benediktbeuern in 1821 when M. Reynier, an assistant of P. L. Guinand, informed the Astronomical Society that Guinand could manufacture flint glass free of bubbles and striae.[123] Reynier sent several of Guinand's flint glass disks from his home in Les Brenets, Switzerland, to London, where Herschel, Davies Gilbert, and William Pearson were asked to judge them.[124] As discussed in greater detail in the next chapter, they concluded that one of disks, a perfect piece of glass nearly 7½ inches (18.5 centimeters) in diameter, was the finest they had ever seen.[125]

The differences between British and Bavarian glassmaking traditions became most pronounced when Herschel attempted to gather information from Fraunhofer on achromatic-lens production on behalf of the Board of Longitude and Royal Society's Joint Committee for the Improvement of Glass for Optical Purposes, founded in April of 1824.[126] In Benediktbeuern, the two traditions and cultures met face to face. During September of 1824, John Herschel traveled to the Continent, where he met up with Fraunhofer on 19 September, in order to obtain information on Fraunhofer's method of optical glass manufacture for the Joint Committee.[127] Herschel, upon being greeted in the former monastery, wrote to his good friend Charles Babbage: "I saw Frauenhofer [*sic*] & all his works but the one most desirable to see: his glass-house, which he keeps enveloped in thick darkness."[128] Clearly, Herschel, like everyone else involved in the attempt to create Fraunhoferian-quality products, thought that witnessing Fraunhofer's workmanship would be the most effective way to glean information on the production of top-grade optical glass. Fraunhofer—lest he give a potential competitor too much insight—would, alas, not permit Herschel to witness the labor practices involved in glass manufacturing, or even the inner structure of the glass hut. Fraunhofer did, however, personally demonstrate to Herschel how he produced the dark lines in the spectrum and how those lines could be used as a system of calibration for achromatic lenses.[129] He revealed nothing else. Herschel was amazed. He continued his letter to Babbage: "One thing is certain, he [Fraunhofer] has completely resolved the problem of making perfect flint glass, homogeneous, [free of striae and veins]."[130] He praised Fraunhofer once again, this time to J. J. Littrow, the professor of astronomy in Ofen, on 9 November 1824:

> I passed some time at Munich with M. Frauenhofer [*sic*] with whom I was much delighted. I will not say that I think his results merely *interesting,* they are admirable & *instinctive.* He is the only optician existing I believe who has aimed at applying strict theory to *exact* data in the construction of his glasses. It is a pity he makes a secret of his glass-making & it were to be wished that he would publish the math. theory and numerical computation of the curvatures of his large achromatics.[131]

Herschel was hinting here that Fraunhofer's mathematics, optical theory, and calculations would enable him to reproduce Fraunhoferian lenses. As will be discussed below, secrecy was antithetical to Herschel's vision of a moral scientific enterprise.

Although he was rather disappointed, Herschel certainly did not despair when he found that Fraunhofer would show him neither the skilled labor involved nor the pieces of glass in different stages of the process, only the finished products.[132] Herschel was rather optimistic at first that the British could ferret out the secrets by using scientific methods. Henry Fox Talbot continued to provide Herschel with lenses from the Optical Institute after Fraunhofer's death in 1826, and Herschel received other samples from the Optical Institute a year later. He wrote to Faraday on 27 August 1827: "I have just got some prisms from Utzschneider's manufactory among which is a large one of flint glass of the utmost perfection being like a piece of solidified water without a trace of a vein or imperfection of any kind. This shews . . . that the problem is capable of being resolved."[133] The resolvable problem to which Herschel was referring was the attempt of the British Joint Committee to produce flint glass as homogeneous as Fraunhofer's. This would suggest that Herschel did not believe that Fraunhofer's personal and practical knowledge was necessary, since superb glass was being produced after Fraunhofer's death.[134] Hence, for Herschel the labor processes were not as dependent upon Fraunhofer as Brewster thought. Herschel also seemed to think that Fraunhofer's superior prisms could be replicated by the Joint Committee without any knowledge of the skilled-labor practices used in Bavaria to produce optical glass, although he certainly never claimed it would be easy or straightforward.

Herschel's trip to Benediktbeuern must also be put in its proper context. He was the leader of scientific reform in Britain in the 1820s and the 1830s. Philosophical reasoning, experimental science, and economic science were, in Herschel's eyes, all interrelated. His insistence that scientific and mathematical knowledge was indeed communicable is not surprising; after all, that is precisely what a Senior Wrangler did: communicate complex technical knowledge. Hence, it now becomes clear why Herschel would claim that if only Fraunhofer would publish his mathematical theory and calculations, he might be able to construct similar lenses and prisms. Herschel's wish to see "the math. theory and numerical computation" is striking since it implies, once again, that artifacts could be reproduced by using mathematical theory where secrecy hid the manual technique. Mathematics was, according to Herschel, universally communicable. When Herschel described the criteria of good flint glass for the subcommittee to imitate, he claimed that they were very desirable, as they

> permit the very ready & simple application of an exact mathematical theory to the construction of object glasses without supposing any knowledge of mathematics in the artist beyond the working [of] the rule of three sums. Frauenhofer [*sic*] constructs all his large glasses from exact measures of their refractive & dispersive powers, and a computation thereon founded on the proper radii to destroy aberrations and till this practice in general [is known,] it is vain to expect perfect glasses, even with good materials.[135]

In Herschel's eyes, the general principles of mathematics, in addition to superior optical glass, were required to reproduce Fraunhoferian lenses. Unlike Brewster (and, as we shall see below, unlike Faraday), Herschel never seemed to appreciate the difficulty of creating optical glass, particularly after Utzschneider demonstrated with his company's prism manufactured after Fraunhofer's death that the knowledge of optical glass production seemed not to be restricted to Fraunhofer.

After his meeting with Fraunhofer in September of 1824, Herschel praised Fraunhofer's experimental technology of using the dark lines in the spectrum as a method of calibrating lenses.[136] Herschel echoed Brewster's sentiment that British optical hegemony was in serious jeopardy:

> I have seen such specimens of Reichenbach & Frauenhofer's workmanship, as has completely cured me of the weakness of national vanity on the point of the higher departments of astronomical instrument making. A school of artists [craftsmen] who have worked with them is gradually rising in various parts of Europe, and without some great effort, British art will no longer hold the prominence it has done.[137]

Herschel continued by emphasizing that Reichenbach and Fraunhofer "make a great mystery" of their work, citing Reichenbach's multiplying index and Fraunhofer's glass. He spoke of Fraunhofer's glass as "faultless," describing one of his prisms as "of large refracting angle, covering the whole aperture of an object glass 4 inches in diameter and giving spectra of the stars *pure* enough to see the dark fixed black lines which cross them. It is indeed as liquefied as the purest water, being absolutely free from veins or inequalities of refraction."[138] Whereas the British were unable to replicate Fraunhofer's glass, Herschel could reproduce the dark lines as a result of Fraunhofer's demonstration. Indeed, Herschel first saw Fraunhofer's dark lines during his visit to Benediktbeuern.

We are told of Herschel's replication of these dark lines (now called the Fraunhofer lines) in Babbage's *Reflexions on the Decline of Science in England*:

> A striking illustration of the fact that an object is frequently not seen, *from not knowing how to see it,* rather than from any defect in the organ of vision, occurred to me some years since, when on a visit at Slough. Conversing with Mr. Herschel on the dark lines seen in the solar spectrum by Fraunhofer, he inquired whether I had seen them; and on my replying in the negative, and expressing a desire to see them, he mentioned the extreme difficulty he had had, even with Fraunhofer's description in his hand and the long time which it had cost him in detecting them. My friend then added, "I will prepare the apparatus, and put you in such a position that they shall be visible, and yet you shall look for them and not find them: after which, while you remain in the same position, I will instruct you *how to see them,* and you shall see them, and not merely wonder [why] you did not see them before, but you shall find it impossible to look at the spectrum without seeing them."
>
> On looking as I was directed, notwithstanding the previous warning, I did *not* see them; and after some time I inquired how they might be seen, when the prediction of Mr. Herschel was completely fulfilled.[139]

Observation of the dark lines was a skill. Fraunhofer had personally demonstrated to Herschel his technique at Benediktbeuern. Even with Fraunhofer's personal description in hand, which he brought back with him to England with several prisms, Herschel needed to teach Babbage how one could produce and "see" the dark lines.[140]

Herschel used the glass crisis to legitimize his own agenda for reform of the Royal Society, but by 1827 he was beginning to realize how difficult that would be. In December he resigned his post as secretary of the Royal Society. After that date, his interest in the subcommittee's glass enterprise waned considerably. More important, the Fraunhofer episode illustrates Herschel's attitude toward skilled artisans. Herschel claimed that the production of lenses could be improved by the tradition of rational and scientific knowledge if opticians and instrument makers were to follow the cue of experimental natural philosophers, who in turn were to render their theories accessible to artisans. One can easily imagine why Brewster, who supported artisanal schools, opposed such a stance.

In a telling letter to Herschel, Robert Fot, a wright and turner who was studying optics at a school of arts patronized by Herschel, expressed regret that, rather than learning the practical skills of the optical trade, such as optical glass production, he was taught Herschel's geometrical optics:

> You in your neighbourhood patronised a school of arts which instance of condescension to the people of the working class, to which I belong[,] emboldened me to solicit the privelege [*sic*] of communicating at times with you. I have not been bred an optician, being by profession a wright and turner, but the whole of my leisure is devoted to the study of optics in so far as the construction of achromatic telescopes are concerned. In all the object glasses I have made, your most valuable formula has always been used. . . . However when after all my pains I fail in making a good one I can manage to better it by a method of polishing anew one side of the Crown lens. But there is one great difficulty continually in my way. I can by no means procure good glass, especially flint.[141]

This letter reflects the general inadequacy of the schools of arts and the Mechanics' Institutes throughout Great Britain. Rather than teach mechanics and artisans the application of their trade to the various forms of manufacture, as Brewster wanted, they provided them with information that was not always relevant or helpful.[142] This sentiment was echoed in a letter from a disappointed mechanic which was published in *The Mechanics Weekly Journal* in December of 1823:

> Scientific lecturers attend to generalities only, and omit the details of practice, with which indeed they are unacquainted. Of course, their lectures (unless rendered unintelligible by the use of philosophical slang, understood only by those who have studied it,) do indeed amuse the hearers, by exhibiting the causes of the effects

observable in the operations of the workshop; but they do not directly contribute to the improvement of the traders dependent upon those principles. This improvement can only be made by those who are practically conversant in the minutest details,—who have acquired, by continual practice, that slight of hand which is so essential to a true performance of any mechanical operation, and who are well acquainted with the difficulties and defects which are to be got over. . . . A large society of this kind [The London Mechanics' Institute] is likely to hire lecturers high in scientific renown, but little acquainted with the application of their knowledge to the arts of life, or the real practice of the workshop, or the commercial laboratory. So that, while their extended views gratify the highly educated few, the operative mechanic, who is drawn in by fair promises to subscribe his mite to their support, finds his object neglected, sinks at the continual mistakes, in minute points, of the learned professor,—withdraws in disgust, and the institution is diverted from its original purpose.[143]

Herschel strongly upheld the notion of the universality, rather than an enculturated model, of scientific skill. His *Preliminary Discourse on the Study of Natural Philosophy*, published in 1830, reveals his belief in the universality of the scientific enterprise: "It is not one of the least advantages of these pursuits [the study of natural science], . . . that they are altogether independent of external circumstances."[144] He continues by extending the notion of the universality of natural science to the universality of knowledge in general:

It [knowledge] acquires not, perhaps a greater certainty, but at least a confirmed authority and a probable duration, by universal assent. . . . Those who admire and love knowledge for its own sake ought to wish to see its elements made accessible to all, were it only that they may be the more thoroughly examined into, and more effectually developed in their consequences, and receive that ductility and plastic quality which the pressure of minds of all descriptions, constantly moulding them to their purposes, can alone bestow. *But to this end it is necessary that it should be divested, as far as possible, of artificial difficulties, and stripped of all such technicalities as tend to place it in the light of a craft and a mystery, inaccessible without a kind of apprenticeship*. . . . Everything that tends to clothe it [science] in a strange and repulsive garb . . . assumes an unnecessary guise of profundity and obscurity.[145]

He stated that practical mechanics, or the domain of knowledge that instructs natural philosophers on how to combine and apply nature's forces for human purposes, belonged to the domain of purely theoretical knowledge:

Practical Mechanics is, in the most pre-eminent sense, *a scientific art;* and it may be truly asserted, that almost all the great combinations of modern mechanism, and many of its refinements and nicer improvements, are creations of pure intellect, grounding its exertion upon a moderate number of very elementary propositions in theoretical mechanics and geometry.[146]

Finally, he criticized "empirical art" since it tended toward incommunicable knowledge. Science, by Herschel's definition, was communicable either by the experimental natural philosophers themselves or by their artifacts. Scientific art, unlike empirical art, made the technical mystery accessible to all cultures:

They [the arts, such as practical mechanics] cannot be perfected till their whole processes are laid open, and their language simplified and rendered universally intelligible. Art is the application of knowledge to a practical end. If the knowledge be merely accumulated experience, the art is *empirical;* but if it be experience reasoned upon and brought under general principles, it assumes a higher character, and becomes a *scientific art*. . . . The whole tendency of empirical art, is to bury itself in technicalities, and to place its pride in particular short cuts and mysteries known only to adepts; to surprise and astonish by results, but conceal processes. The character of science is the direct contrary. It delights to lay itself open to enquiry; and is not satisfied with its conclusions, till it can make the road to them broad and beaten: and in its applications it preserves the same character; its whole aim being to strip away all technical mystery, to illuminate every dark recess, and to gain free access to all processes, with a view to improve them on rational principles. It would seem that a union of two qualities almost opposite to each other—a going forth of the thoughts in two directions, and a sudden transfer of ideas from a remote station in one to an equally distant one in the other—is required to start the first idea of *applying science*.[147]

Herschel was disappointed by Fraunhofer's secrecy, which he claimed was anathema to "the character of science." Although secrets and craft knowledge are two distinct entities, they were very intimately related during the early nineteenth century. In stark contrast to Brewster's support of inventor's secrecy, Herschel called for an opening of artisanal secrets on the ground that they could not be easily deduced by abstract science. But Fraunhofer, according to Herschel, must have practiced scientific art; after all, Herschel awarded Fraunhofer an associate membership in the Astronomical Society of London in 1825, and he labeled the Bavarian both an artist and a philosopher.[148] Hence, he thought he could learn Fraunhofer's procedures for glassmaking during his visit in 1824. Failing that, he thought that Faraday would be able to recover the procedures by "reverse engineering" the final product. The transfer of technology from Benediktbeuern to London is precisely what Herschel meant by "applying science." But by 1827, after two years of failing to replicate Fraunhoferian glass, Herschel began to despair. In his minutes of the General Scientific Committee he claimed that "the character of science is itself in some degree at stake" should Faraday, George Dollond, and he fail at this venture.[149] The Joint

Committee, in Herschel's view, needed to succeed; otherwise his implication that the specific techniques are irrelevant, that generic techniques can be obtained by applying general principles, would be undermined. This was Herschel's definition of "the character of science." Herschel's belief in the universality of science is also evident from his belief that the rise of new journals throughout Europe placed "observers of all countries on the same level of perfect intimacy with each other's objects and methods" and "serve to direct the course of general observation, as well as to hold out, in the most conspicuous manner, models for emulative imitation."[150]

Herschel was attracted to Fraunhofer's work because of the precision measurements that the Bavarian's instruments produced. In his *Preliminary Discourse* he noted the importance of improved observation by means of instruments adapted for exact measurement: "What an important influence may be exercised over the progress of a single branch of science by the invention of a ready and convenient mode of executing a definite measurement."[151] The crucial difference between Herschel's and Brewster's views is that Herschel claimed that the general scientific principles, which are communicable, could imitate the skilled labor of artisans by deducing the requisite skills from a finished product. And, whereas Brewster deemed trade and guild secrets necessary for artisans, Herschel saw such clandestine work as impeding the progress of science.

Herschel's solution to Britain's dilemma was quite different from Brewster's. Herschel stated that reflecting telescopes, and not refracting telescopes or telescopes with liquid lenses, were the answer to the problem of achromaticity. In the 1820s he committed himself completely to the instrument his father had perfected: the astronomical reflector.[152] His star catalogues of the 1820s and the 1830s and his famous *Cape Observations* of 1847 were compilations of his observations through a reflecting telescope.

After Fraunhofer's construction of the Dorpat refractor, Struve became Herschel's main rival. Indeed, it was claimed that Fraunhofer's refractor was as good as a reflecting telescope with twice the aperture.[153] Talbot inquired of Herschel whether he had read the "remarkably interesting paper by Struve in no. 72 of *Astronomische Nachrichten* in which he gives a list of 113 new double stars discovered by him with the great achromatic [made by Fraunhofer] in a space where only 40 were known before."[154] Indeed, Struve found 3063 double stars from the first through the fourth classes with Fraunhofer's refractor, while Herschel and South had earlier found

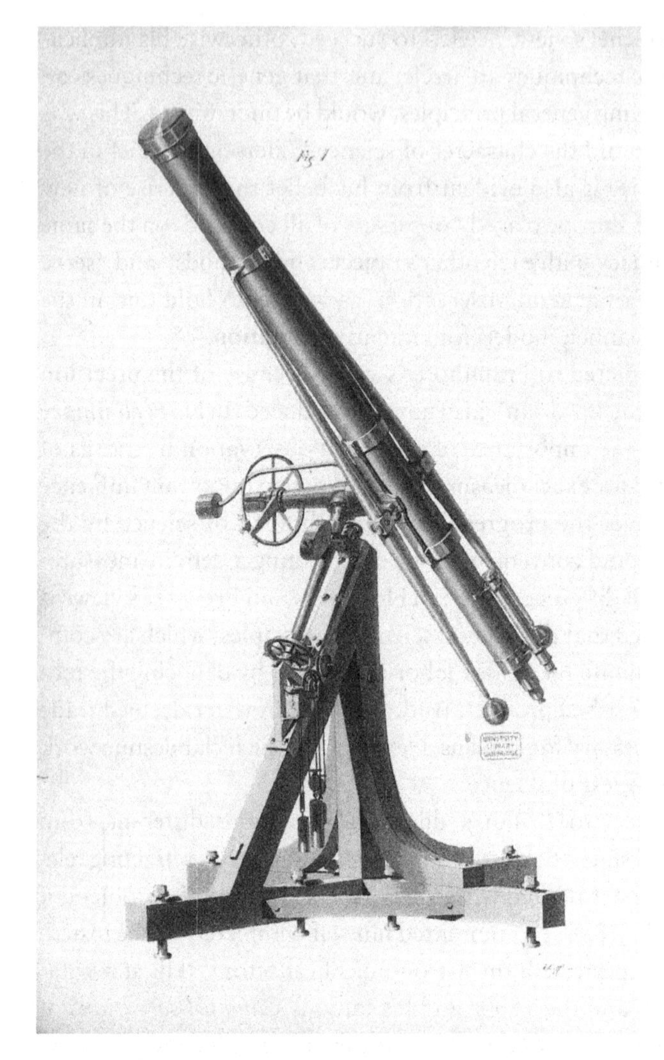

Figure 5.3
Fraunhofer's refractor at Dorpat, used by Friedrich Georg Wilhelm Struve from 1825 on. *(Syndics of Cambridge University Library)*

only 340.[155] Struve triumphantly proclaimed: "We may perhaps rank this enormous instrument with the most celebrated of all reflecting telescopes, viz. [William] Herschel's; whilst it surpasses it in its convenience for use, and the variety of applications. Thus I am inclined to consider our achromatic refractor as the most perfect optical instrument yet in existence."[156]

Herschel became quite angry at the suggestion that Fraunhofer's refracting telescopes were superior to the reflectors manufactured by his father. Fraunhofer himself suggested that refractors were inherently more precise instruments than reflecting telescopes. He argued that the mirrors of a reflecting telescope reflect only a small portion of the light it receives, thereby limiting the intensity of light seen by the observer. Also, he claimed that it was impossible to correct the deviation of the rays occasioned by the spherical form of the reflecting surfaces. Such a deviation greatly hindered resolution under high magnification.[157] Finally, Fraunhofer argued that refractors were superior because glass lenses, made from Bavarian flint and crown glass, could correct both chromatic and spherical aberration. He then took the opportunity to emphasize the superiority of Bavarian artisanal work over British:

> The English flint-glass has undular lines which disperse the light irregularly in its passage through it. . . . The English crown-glass too, as in fact every other kind of glass hitherto used, has those undular streaks, which, although not always visible to the naked eye, will give a false direction to the rays by an irregular refraction. The Bavarian flint- and crown-glass, however, is free from these streaks, and equally compact throughout: the difference between the flint- and crown-glass being chiefly in the greater power of dispersing the colours.[158]

Herschel responded to Fraunhofer's claim by defending his father's telescopes in a cordial letter to Schumacher to be published in *Astronomische Nachrichten*. He starts his letter by praising Fraunhofer's achromatic refractor at Dorpat as "probably the best refracting telescope ever made."[159] He continued, however, "to obviate an erroneous impression which may arise in the minds of those who read Mr. Fraunhofer's memoir, as to the *great inferiority* of reflecting telescopes in point of optical power, to achromatics in general."[160] He rebutted Fraunhofer's assertion that the loss of light owing to metallic reflection was very great, thereby greatly decreasing the amount of light that enters the eye. Herschel argued that metallic mirrors reflect 0.673 of the incidental light and claimed that Fraunhofer's value of 0.452 was based on Newtonian reflectors that used metallic mirrors rather than one.[161] Herschel concluded the letter in a typically tactful manner.

Indeed, the letter as a whole serves as a fine example of how a gentleman can dispute a colleague's claim without appearing to challenge his character:

> I should be very sorry to have expressed myself in any way capable of being construed in a controversial sense, or as intended to give the slightest personal offence to its celebrated author, who as an artist must surely be ever regarded as a benefactor to astronomy, while optical science is no less indebted to him as a philosopher, for his beautifully delicate experiments on the constitution of the prismatic spectrum, which have given a degree of precision to optical determinations hitherto unheard of, and shewn the practicability of placing the construction of telescopes on purely scientific grounds, while they have unfolded phenomena of the highest interest in a speculative point of view. Nor can I help feeling that I should ill requite his liberal and friendly reception during a visit to Munich . . . by a word calculated to give pain or excite unpleasant feelings.[162]

James South backed his compatriot by writing to Schumacher that he was "far from entertaining the sentiments of Mr. Fraunhofer as to the decided superiority of refractors over reflectors." He also disagreed with Struve's assertion that the Dorpat telescope "rank[s] with the most celebrated of all reflecting telescopes, namely, Herschel's."[163] In the 1840s, Lord Rosse of Ireland attempted to salvage British honor in astronomy by means of astronomical observations with a reflecting telescope. Earlier he had commented: "Many practical men to whom I have spoken seem to think that after Fraunhofer's discoveries, the refractor has entirely superseded the reflector, and that all attempts to improve the latter instrument are useless."[164]

Herschel and (to a much greater extent) Babbage belonged to a tradition, nearly 100 years old, that criticized craft knowledge for its secrecy. Denis Diderot, the editor of the *Encyclopédie*, espoused such a belief in Enlightenment France. He wished to free the mechanical arts from the "ignorance and deception" of laborers by opening up craft knowledge. Although he argued both for the importance of artisanal labor and the inability of theoretical knowledge to explain craft processes, Diderot abhorred artisanal corporations, because their secrecy thwarted innovation as well as attempts by scientific knowledge to penetrate and explain artisanal practices. Once these labor processes were revealed, Diderot argued, they could be properly guided by enlightened managers.[165] Babbage himself spent more than 20 years having skilled artisans attempt to build his difference engine, which (rather ironically) was meant to replace the labor of human calculators.[166] One of the models for his intelligent calculating engine was Jean Jacquard's weaving looms, which were spreading through Paris during the first decade of the nineteenth century.[167] Although

Herschel was not quite as extreme as Babbage, he did not think that artisanal knowledge could itself be compared in character to natural philosophical knowledge, and he reckoned that the rational principles of optics and mechanics could suffice for the production of artifacts such as lenses and prisms.

The concept of skill was the point of contention between Herschel and Babbage, on the one hand, and Brewster, on the other. According to Brewster, skill was a property inherent in the skilled artisans, or inventors, themselves. It could be communicated only in a controlled milieu, such as at a workbench or in a laboratory. Skill was to remain hidden from the surveillance of managers.[168] Babbage and other philosophers of machinery supported an account of rational valuation that attempted to render the skills of artisans visible, quantifiable, and manageable in the marketplace of wage labor.[169] As Ashworth (1996) convincingly demonstrates, both Herschel and Babbage were fundamentally concerned with the efficiency of industrial manufacture and with how such efficiency could be applied to the human mind. Both studied British factories, such as Boulton and Watt's Foundry, in order to observe, first hand, organizational schemes of industrial labor. Ashworth concludes: "To Herschel, Watt's mind seemed to be organized exactly like the Soho Foundry: to view the foundry was to see a great mind. The emphasis was on systematic and efficient access to correctly recorded and processed thoughts."[170] Herschel did his best to render what he referred to as the "technical mystery" of Fraunhofer's skills visible and quantifiable in order to increase British production of optical lenses, while Brewter argued that attempts to penetrate Fraunhofer's highly secretive laboratory were futile.

Both Brewster and Herschel were fundamentally concerned with the role of artisans within the nation's branches of manufacture. That is one of the major reasons why the Fraunhofer episode intrigued them both so much. But Herschel and Brewster differed on how the mechanical skill of artisans should be dealt with in terms of manufacture. For Brewster, the "practical knowledge" and "mechanical genius" of skilled artisans were crucial for Britain's recovery from its scientific and economic decline. His bête noire, the British government, needed to revamp its patent laws in order to secure the rights of inventors and thwart attempts of dishonest artisans and "capitalists," who shamelessly sought the financial rewards deserved by others. Because British legislation failed to achieve such safeguards, secrecy was a legitimate form of social behavior; it was, for Brewster, a means to an end. In his portion of the 1830 essay co-authored with Babbage, Brewster concluded that one reason "the sciences and arts of England [were] in a wretched state of depression" was the "unjust and oppressive tribute which the patent-law exacts from inventors."[171] Since skill was the property of the inventor, the mechanical genius, Mechanics' Institutes should recognize such skill and instruct the artisans in applying their talent to manufacture for their own "moral good," as well as for the benefit of the public and the nation.

Herschel, on the other hand, like Babbage, was interested in artisanal knowledge in order to quantify and manage it. Once it was stripped of its craft secrecy, it could be efficiently managed by and applied to industrial manufacture. The "technical mystery" of craft knowledge impeded such efficiency and, in Herschel's eyes, threatened Britain's economy. It also stood in sharp contrast to the openness of science.

Michael Faraday

Michael Faraday (1791–1867)[172] was born to a devoutly Sandemanian working-class family in what is now Southwark, London. His father, a blacksmith in poor health, died when Michael was 18. Michael was apprenticed to a bookbinder and became fascinated with texts on experimental natural philosophy, particularly the article on electricity in the *Encyclopaedia Britannica*. In 1812 Faraday went to listen to the public lectures of Sir Humphry Davy at the Royal Institution. In late October of that same year, Davy was temporarily blinded by a chemical explosion in his laboratory. Faraday was recommended to Davy as an amanuensis. This meeting changed Faraday's life. In December, once Davy regained his sight, Faraday sent Davy the carefully bound notes he had taken at the public lectures. In February of 1813, when Davy's assistant was fired for brawling, Davy sent for Faraday as a replacement. On 1 March 1813, Faraday took up his new position as Davy's assistant at the Royal Institution.

Davy taught the young Faraday a great deal of chemistry. The two traveled together to France and Italy during Davy's lecture and research trips in 1813 and 1814. By 1820 Faraday had established himself as a very competent analyst and chemist. He also showed an aptitude for applying chemistry to technological problems such as the production of high-grade steels. During the 1820s and the 1830s, Faraday undertook promising research

Figure 5.4
Michael Faraday (1791–1867). *(Deutsches Museum, Munich)*

on electricity, particularly electromagnetic rotation (in 1821), electromagnetic induction (in 1831), and the laws of electrochemistry. In 1825 the Joint Committee of the Board of Longitude and the Royal Society asked Faraday to experiment with optical glass in order to improve its manufacture in Britain.

Faraday, as we will see, attempted to replicate Fraunhoferian glass by chemical means, and in the process invented a new glass technology. He drew upon the chemical knowledge he had gathered under Davy and continued William Thomas Brande's[173] research on new types of glass made from the borate of lead and the borate of lead with silica. Several glass samples composed of the silicated borate of lead, he found, could be used rather than flint glass in a telescope.[174]

Faraday, by virtue of his employment at the Royal Institution, was concerned with the communicability of scientific knowledge. The Royal Institution was founded with the purpose of "diffusing the Knowledge, and facilitating the general Introduction, of Useful Mechanical Inventions and Improvements; and for teaching, by courses of Philosophical Lectures and Experiments, the application of Science to the common Purposes of Life."[175] Beginning in early 1826, a year after being approached by the Joint Committee, Faraday began to expand the Royal Institution's program to disseminate scientific knowledge to a larger audience, the educated general public.[176] He wished to communicate the most abstract and technical knowledge of the laboratory to a wider audience.[177] It is in this context that his rarely cited *Chemical Manipulation* (1827) should be considered. Although the book deals exclusively with the performance of experiments, or manipulation, Faraday claims that manipulation has a "subordinate character" in respect to the scientific principles upon which the philosopher uses to devise the experiment.[178] He continues, however, by arguing as follows:

> Notwithstanding this subordinate character of manipulation, it is yet of high importance in an experimental science, and particularly Chemistry. The person who could devise only, without knowing how to perform, would not be able to extend his knowledge far, or make it useful. . . .The importance of instruction in manipulation has long been felt by the author during his professional experience as a public and private teacher of Chemistry in the Royal Institution; and the deficiency existing in the means of teaching it, induced him to think he might perform an acceptable service by putting together such information on the subject as there was reason to suppose would be generally useful to the student. [179]

Nevertheless, Faraday, renowned for his skilled manipulations, admits that "no book contains those minute directions which are necessary in the present extensively cultivated state of the science, nor can verbal instruction teach that perfection of manipulation which is to be gained by constant operation."[180] He separates the "practice of science" (which he seeks to teach) from "the principles of science" (which, quite explicitly, he fails to inculcate,[181] on the ground that it teaches not "a habit of reasoning" but an "art of experimenting"[182]).

At first Faraday shared Herschel's view of "scientific art" as well as his optimism that Fraunhofer's secrets might be uncovered. For several years Faraday tried to replicate Fraunhofer's optical glass. At first, Faraday thought that to do this he needed only to have samples of Fraunhofer's best glass. Such a view stood in sharp contrast to Brewster's and Fraunhofer's. As time went on, however, Faraday, faced with failure of reproducing Fraunhoferian-quality flint glass, recognized that there might be something missing:

> But be it remembered that it is not a mere analysis, or even the developement of philosophical reasoning that is required: it is the solution of difficulties, which, as

is the case of Guinand and Fraunhofer, required many years of a practical life to effect, if it was ever effected. It is the foundation and developement of a manufacturing process, not in principle only, but through all the difficulties of practice, until it is competent to give constant success.[183]

This quotation is from a paper—which Faraday first presented as a two-part Bakerian Lecture before the Royal Society in late November and early December of 1829—summarizing the results of his experiments on glass. The purpose of this lecture and of the subsequent essay was to describe the process of optical glassmaking so that a "practical man" (glassmaker) could draw upon his experience and improve upon its manufacture.[184] The paper began with a tribute to Fraunhofer's and Guinand's practical knowledge:

It must be well known to the scientific world, that these difficulties [of glass manufacturing for optical purposes] have induced some persons to labour hard and earnestly for years together, in hopes of surmounting them. Guinand was one of these: his means were small, but he deserves the more honour for his perseverance and his success. He commenced the investigation about the year 1784, and died engaged in it in the year 1823. Fraunhofer laboured hard at the solution of the same practical problem. He was a man of profound science, and had all the advantages arising from extensive means and information, both in himself and others. He laboured in the glass-house, the work-shop, and the study, pursuing without deviation the great object he had in view, until science was deprived of him also by death. Both these men, according to the best evidence we can obtain, have produced and left some perfect glass in large pieces.[185]

Faraday continued by pointing out the possibility that "the knowledge they acquired was altogether practical and personal, a matter of minute experience, and not of a nature to be communicated."[186] When Faraday's own manipulative skills proved insufficient, he decided that the concrete, skilled process of optical glass production could not be obtained from glass itself. By 1828, Faraday had concluded that the English could simply not produce flint glass as well as their Bavarian counterparts, and he turned his attention to heavy glass with the silicated borate of lead. His attempts at reverse engineering failed.

The central theme of this chapter has been the response to Fraunhofer's superb artisanal skills and practices of glassmaking by leading British experimental natural philosophers. The skill and labor involved in artisanal practices provide a heuristic tool for analyzing how these philosophers formulated the reworking of physics within a new reformed society. During the 1820s, Herschel, Babbage, Brewster, and Faraday defined the context of physics within a rapidly changing society at a time when the nature and the role of skill and practice were at stake in both the laboratory and the workplace. As John Rule has demonstrated, "the sense of a property of skill was . . . deeply embedded in the culture and consciousness of the artisan, as was the assumption of the respect of others for it."[187] Skill, in a sense, was the only property that artisans could claim as their own. It was creative and inventive. William Cooper, a Scottish maker of crown glass, argued strenuously and passionately that "mechanical skill" was what granted pride to the artisan and elevated both his "character" and his "trade":

> The general diffusion of knowledge, and the improvements in the modes of its dissemination which have occurred within these few years, have made literature as familiar in the workshop of the artisan as in the closet of the student. . . . This association of literature with mechanical skill must have a beneficial effect in two respects. In the first place, its tendency is to raise the humblest mechanical trades to the rank of the sciences, and impart to the workman a sense of the worth and importance of his art, and such a knowledge, as well of its theory as its practice, as every craftsman ought to possess. It will elevate his own character by elevating that of his art, and, while it promotes his general intellectual advancement, it will, in the second place, enable him to refine and improve on that art to an extent which must soon be felt in the comforts, elegancies, and conveniences of life. We cannot but think that the skill and ingenuity of the mechanic, in ordinary things, are not sufficiently prized, and this merely because, from not being sufficiently known, they are accounted ordinary things. We look with indifference on the produce of his intellect and the works of his hands, and say, with an emphasis which is meant to deprive him of all merit, "It is his trade." True, it may be his trade, but what of that? Has there been no mind at work to regulate, no genius to direct his manual dexterity? Yes; in the meanest mechanical art there is much of these qualities displayed. Genius has, indeed, been allowed to the tradesman, and this quality of mind has been denominated mechanical genius.[188]

By using the term "mechanical genius," Cooper was arguing that mechanical skill was both creative and inventive. It was not, as Edward Young had claimed nearly 100 years earlier, a simple reworking of what already existed. Cooper added:

> Now, whilst this term [i.e., "mechanical genius"] is understood merely in a discriminative sense, to distinguish it from the other forms under which genius appears, and when it is not associated with any ideas of inferiority, it is well enough; but when it is understood as coupled with certain ideas of inferiority, we cannot assent to the propriety of that disparagement of the faculty which is thus implied; for it cannot be denied, that genius exercised in making a shoe, or in any other piece of workmanship, is as distinctly and purely genius as that faculty developed in writing a book. It is true, that in the former of these works it appears in a less dignified form than in the latter, but is its result less an emanation of genius on that

account? A diamond is a diamond still, whether it be placed on the blue bonnet of the peasant, or the monarch's crown; and who knows but that genius which may have discovered itself in the humble and homely form spoken of, might have raised its possessor to a high place amongst men, had his destiny been otherwise ordered? Or who can say, that what is called mechanical genius, may not often be an accident of the quality, the result of a chance direction of its powers; thus making it possible, that Watt and Bonaparte might have exchanged places, without loss of any of that fame, which we now associate with the names of these extraordinary men. Apart from these considerations, however, who does not see, that the discoveries of Watt have been infinitely more valuable to mankind, and are attended with a much purer glory, than the matchless but desolating conquests of Bonaparte?[189]

Cooper makes two critical points in the second half of this lengthy quotation. First, "mechanical genius" is a term to be used in contrast to works of less skilled artisans, or even unskilled laborers. Second, and more important, Cooper is strongly intimating that society defines genius, and its associated skill and creativity, in respect to the class of persons who possess it. Hence, genius is affiliated with writing a book, but never with making a shoe. He feels that the gift of genius is bestowed on individuals from all walks of life, and that the products of genius are equal whether they are found on a peasant or on a monarch. Unfortunately, in the 1820s and the 1830s, within certain circles of British society—circles that included several leading experimental natural philosophers—the skill of the mechanical genius was seen as subordinate to intellectual genius.

6

The British Response to the Bavarian Threat

Fraunhofer's calibration technique of using the dark lines of the spectrum for determining refractive indices rapidly became the standard in optical-lens production. In 1825, Brewster wrote that it was necessary to turn to Fraunhofer's essay on the Six Lamps Experiment for the "most accurate knowledge" about the calibration of achromatic lenses.[1] He even defined "an excellent prism" as one "capable of showing Fraunhofer's lines."[2]

In his 1831 *Treatise on Optics*, while explaining Newton's analysis of the colors of the spectrum, Brewster confessed that when English glass prisms were used "no lines are seen across the spectrum . . . , and it is extremely difficult for the sharpest eye to point out the boundary of the different colours" while replicating Newton's experimentum crucis.[3] He then offered a table that compared the lengths of the colors of the spectrum as experimentally determined by Newton against Fraunhofer's results obtained by using the Bavarian's and Guinand's flint glass (figure 6.1).[4] The old standard-bearer had been replaced by a new, foreign one. Later in the treatise Brewster laments:

> in consequence of the great difficulty of obtaining flint glass free of veins and imperfections, the largest achromatic object glasses constructed in England did not greatly exceed 4 or 5 inches in diameter. The neglect into which this important branch of our national manufactures was allowed to fall by the ignorance and supineness of the British government, stimulated foreigners to rival us in the manufacture of achromatic telescopes. M. Guinand of Brenetz, in Switzerland, and M. Fraunhofer, of Munich, successively devoted their minds to the subject of making large lenses of flint glass, and both of them succeeded.[5]

By the 1850s, British optical texts routinely began with brief histories of their subject. Fraunhofer's work on the dark lines of the solar spectrum was lengths of the colours to be as follows, in the kind glass of which his prism was made. We have add the results obtained by Fraunhofer with flint glass.

	Newton.	Fraunhofer.
Red - -	45	56
Orange - -	27	27
Yellow - -	40	27
Green - -	60	46
Blue - -	60	48
Indigo - -	48	47
Violet - -	80	109
Total length	360	360

These colours are not equally brilliant. At the lowe end, L, of the spectrum the red is comparatively fain but grows brighter as it approaches the orange. Th light increases gradually to the middle of the yellov space, where it is brightest; and from this it graduall declines to the upper or violet end, K, of the spectrum where it is extremely faint.

(64.) From the phenomena which we have now de scribed, sir Isaac Newton concluded that the beam o white light, S, is compounded of light of seven differen colours, and that for each of these different kinds o light the glass, of which his prism was made, had different indices of refraction; the index of refraction fo the *red* light being the least, and that of the violet th greatest.

Figure 6.1
Brewster's (1831a) comparison of a Fraunhofer prism with one produced by Newton.

their common point of origin. For example, William Simms's work on achromatic telescopes, published in 1852, states:

In former times, much uncertainty attended all experiments on the solar spectrum with a view to the correction of colour in consequence of the difficulty of defining the limits of each colour. They are so softened off and blended one with another, that it is impossible to determine, with any degree of certainty, where one ends and another begins. A discovery [of Fraunhofer's] of modern times, however, has removed this difficulty. The spectrum is crossed by dark lines, visible through a telescope, which divide its length into definite series. These lines are distinguished by a few of the letters of the alphabet as A, B, C, etc.; and it is now usual, when anything like precision is aimed at, to express the dispersive powers of media by the lengths of the spaces included between the lines A and B, B and C, and so of the rest.[6]

By the middle of the nineteenth century, Fraunhofer's technique of depicting the dark lines of the solar spectrum had become routine for opticians throughout Europe. Fraunhofer's calibration technique had become well established, meaning that Fraunhoferian-quality glass was by that time widely available.

The Joint Committee's and the Subcommittee's Research on Achromatic Flint Glass

From the mid 1810s on, Utzschneider and Fraunhofer's Optical Institute received scores of orders from around the world. As was briefly mentioned in the previous chapter, the British response to the Bavarians' ever-extending lead was the creation of the Joint Committee of the Board of Longitude and the Royal Society for the Improvement of Glass for Optical Purposes on 1 April 1824[7]:

> The President of the Royal Society [Humphry Davy] having observed that the present state of the glass manufactured for optical purposes was extremely imperfect, and required some public interference: it was resolved to require the President and Council of the Royal Society to appoint a Committee to confer with the resident Members respecting the best mode of conducting such experiments as they may think necessary at the expense of the Board.[8]

The Joint Committee intended explicitly to produce glass that would rival Fraunhofer's.[9] The members included such scientific luminaries as Humphry Davy, Davies Gilbert, Thomas Young, John Herschel, and William Hyde Wollaston, none of whom was familiar with the artisanal techniques of glassmaking.[10] At the first meeting of the Joint Committee, on 20 May 1824, the London optician George Dollond was asked to inquire of the glassmakers Apsley Pellatt Jr. and his son-in-law Green of the Falcon Glass Works in London as to the cost "of erecting a furnace capable of producing a very high temperature for experiments upon the manufacture of different kinds of glass."[11] In a letter to Thomas Young dated 25 May 1824, Pellatt and Green responded to the Joint Committee's query:

> We shall consider it an honour if we can contribute in any way to so desirable an object and feel willing to let them a piece of ground adjacent to our premises in Castle Yard[,] Holland Street [in Black Friars] which has a small house attached (the latter may be fitted up as a laboratory) free of any charge whatever. . . . We should think a small furnace with a kiln for annealing sufficient to contain three small Pots of 3 c.w.t. each would cost about £200 to £300. In the event of the furnace

Figure 6.2
Royal Society bust of George Dollond, by Garland. *(President and Fellows of the Royal Society of London)*

being built, if required, we shall be happy to superintend the same to supply pots and materials necessary for glass making on payment of our expenses.[12]

The furnace was semicircular in design, 6 feet high, with a chimney that measured nearly 45 feet. A cave was constructed underneath the chimney to create a draft. It was designed to hold three pots, each pot having a maximum load of 1½ hundredweight of glass.[13] It was a typical British glasshouse: small (21 feet by 15 feet), but very high. These coned-shaped houses, quite famously, dotted the landscapes of eighteenth- and nineteenth-century British industrialized cities. The large coned chimneys created forceful drafts that would increase the air circulation through the burning coals, thereby increasing the temperature of the molten glass.

As the furnace was coming to completion, Pellatt had one major concern on his mind: the excise duties. He wrote to Thomas Young stating that he had been contemplating

> the difficulty you will experience in procuring your coals & getting rid of the Cinders in the experimental furnace unless the Treasury oblige the Excise to afford every accommodation by permitting the small furnace to communicate with our coal yard & premises. Now would be the time to have the thing determined before we complete the door ways, communications & c. I should therefore suggest to you & the Committee to procure an appointment with the Treasury or excise.[14]

Young petitioned the commissioner of excise, who permitted that "experiments may be allowed to be performed without payment of Duties . . . and [the Joint Committee] may . . . conduct their experiments without the interference of the officers of the Department of the Revenue."[15] The commissioner of excise, seeing the importance of this research to the British scientific community, agreed to exempt the Joint Committee's research from the duty.

Pellatt and Green's furnace was completed by 5 May 1825, and experiments on glass manufacture (often supervised by Dollond and Herschel) commenced on 31 May.[16] Pellatt and Green attempted to produce optical flint and crown glass, following the recipes Faraday had ascertained from samples of optical glass in his possession, such as Guinand's and Fraunhofer's.[17] In their first experiment, they varied the ratios of silica, lead oxide, pearl ash, and potash for flint glass.[18] But problems immediately arose. The clay pots that held the flint glass either cracked or reacted with the glass' ingredients under the intense heat.[19] The flint glass in particular took its toll as the lead oxide often dissolved the pot or crucible.

At the urging of Herschel, the Joint Committee established a subcommittee to oversee Pellatt and Green's experiments. This subcommittee, created on 5 May 1825, was to reverse engineer Guinand's, and later Fraunhofer's, lenses[20]; it consisted of Herschel, Dollond, and Faraday. Faraday investigated the chemistry of glassmaking. Dollond worked on the glass and rendered his judgment on its use as an object lens, while Herschel examined the physical properties of the optical glass.[21] Despite the collaboration with the subcommittee, Pellatt and Green could not produce homogeneous flint glass; striae and bubbles were present throughout the glass blanks.

Figure 6.3
Eighteenth-century illustration of an English conical glasshouse. *(Science & Society Picture Library of the Science Museum, London)*

Although a glass blank might appear to be striae-free to the naked eye, it still might not be homogeneous, and Herschel knew that the degree of homogeneity had to be tested thoroughly. Indeed, as was noted in chapter 2, he drafted a list of qualities of Fraunhoferian flint glass which he thought the British should imitate: perfect transparency, homogeneity of refractive indices, and a dispersive power between 1.667 and 2.000 (if ordinary optical crown glass is defined as 1.000).[22]

Homogeneity proved to be the most difficult problem for the British. The technique Faraday employed to test a glass blank's homogeneity was the measurement of its specific gravity at the top, the middle, and the bottom.[23] If there was a large variation between the specific gravities of different layers of a glass blank, it would be nearly impossible to correct for chromatic aberration.[24] In a letter to Faraday (dated 13 September 1825) describing

Pellatt and Green's early attempts to manufacture glass blanks, Herschel expressed his amazement over how each sample possessed such a wide range of specific gravities:

> The difference of s.g. [specific gravity] between the top and bottom in one case exceeds anything I could have supposed possible. It is evident that no accidental defect in mixing could have produced it. A separation, by subsidence, of one fluid within the other, has evidently taken place. This is very remarkable, and indicates that we ought to aim at making atomic compounds, or at least compounds *capable* of permanent mixture. I therefore formed an atomic silicate of lead. Its refractive index came out as high as 2.123 for Extreme red rays & therefore possibly as high as 2.2 for [the] mean, which approaches the refraction of phosphorous and exceeds glass of antimony. Its dispersion is enormous—so much so that I could not measure it with my usual apparatus. A prism of 21°12' required to be opposed by three prisms of flint glass of 30° to neutralise the colour.[25]

This passage is revealing as it suggests that Herschel was less interested than Brewster or Faraday in the skilled labor needed to produce homogeneous flint glass. He felt that the answer to the dilemma of making good optical glass was to use the rational principles of mathematics and geometrical optics. His suggestion of lead silicates is also informative. In February of 1826 the Joint Committee decided to draw upon the expertise of William Thomas Brande, who had previously worked with the borate of lead, the silicated borate of lead, and glass with borax.[26] Pellatt and Green were instructed to execute experiments using Brande's ingredients in their proper proportions. So between February 1826 and April 1827, Pellatt and Green, under the watchful eye and direction of the subcommittee, attempted, without success, to produce blanks of flint and crown glass and heavy glass made from the silicated borate of lead and borax.[27] On 17 April 1827, Herschel wrote to Royal Society president Davies Gilbert that the last glass experiment had been a complete failure and that it was "absolutely impossible for the glass committee to do anything effectual without a *regular operator* to be on the spot and superintend the processes."[28] Dollond had given all the attention he could to Pellatt and Green, and Herschel occasionally gave general instructions on the course of experimentation. Faraday "gave no assistance at all."[29]

Herschel was beginning to become rather impatient. Because "so high an interest [was] excited" by the work of the Joint Committee, and (as was mentioned in the previous chapter) "the character of science itself [was] at stake," Herschel underscored "that nothing loose, indistinct, or irregular, either in points of management, or statements, should occur."[30]

He concluded by arguing that one major reason for the failure thus far was that Pellatt and Green, as ordinary glassmakers unfamiliar with the practices of making optical glass, did not possess the requisite knowledge of optical glass manufacture: ". . . the objects in view [to produce optical glass] have required much, & will require more deviation from the ordinary operations practiced in the manufacture of glass for commercial purposes & a degree of refinement neither obtained nor desired in these [commercial] operations."[31]

Since the Falcon Glass Works, where Pellatt and Green worked, was nearly 3 miles from Faraday's place of employment, the Royal Institution, it was not convenient for him to perform the necessary experiments. Hence, on 8 May 1827, the subcommittee recommended that the Joint Committee formally request the managers of the Royal Institution to erect, at its own expense, a furnace to carry out the glass experiments.[32] The Institution agreed to build a furnace in a small room adjacent to the Battery Room. It also decided to make a portion of the Battery Room accessible to the subcommittee for mixing the ingredients, but it did not agree to finance that portion of the project. The building of the furnace began in May, when the subcommittee approved a plan proposed by the London bricklayer Ramsey at the cost of £115.10.[33] In September of 1827, Faraday hired Sergeant Charles Anderson of the Royal Artillery to assist him.[34]

By the autumn of 1827, the subcommittee was under the control of the General Scientific Committee of the Royal Society. On 6 November, Herschel detailed the general protocol of the subcommittee for the scientific committee. All experimental design was put under Faraday's jurisdiction. Herschel informed him that "with regard to the train of experiments you may think is necessary to engage in, Mr Dollond I am sure, as well as myself, feel every disposition to defer to your superior chemical knowledge."[35] Dollond was named the chairman, Faraday the journalist and treasurer, and Herschel, the secretary. The subcommittee was to keep a regular journal that detailed all the experiments conducted and any of their alterations. In addition, a book was to be kept in which anyone could enter suggestions on any experiment to be considered by the subcommittee. The subcommittee was to make three reports per annum: one after Christmas, one after Easter, and one annual summary to be presented at the first council after the meeting of the Royal Society each November.[36]

In response to the continued failure of the glassmakers, Faraday was requested to play a more active role within the subcommittee.[37] Rather than

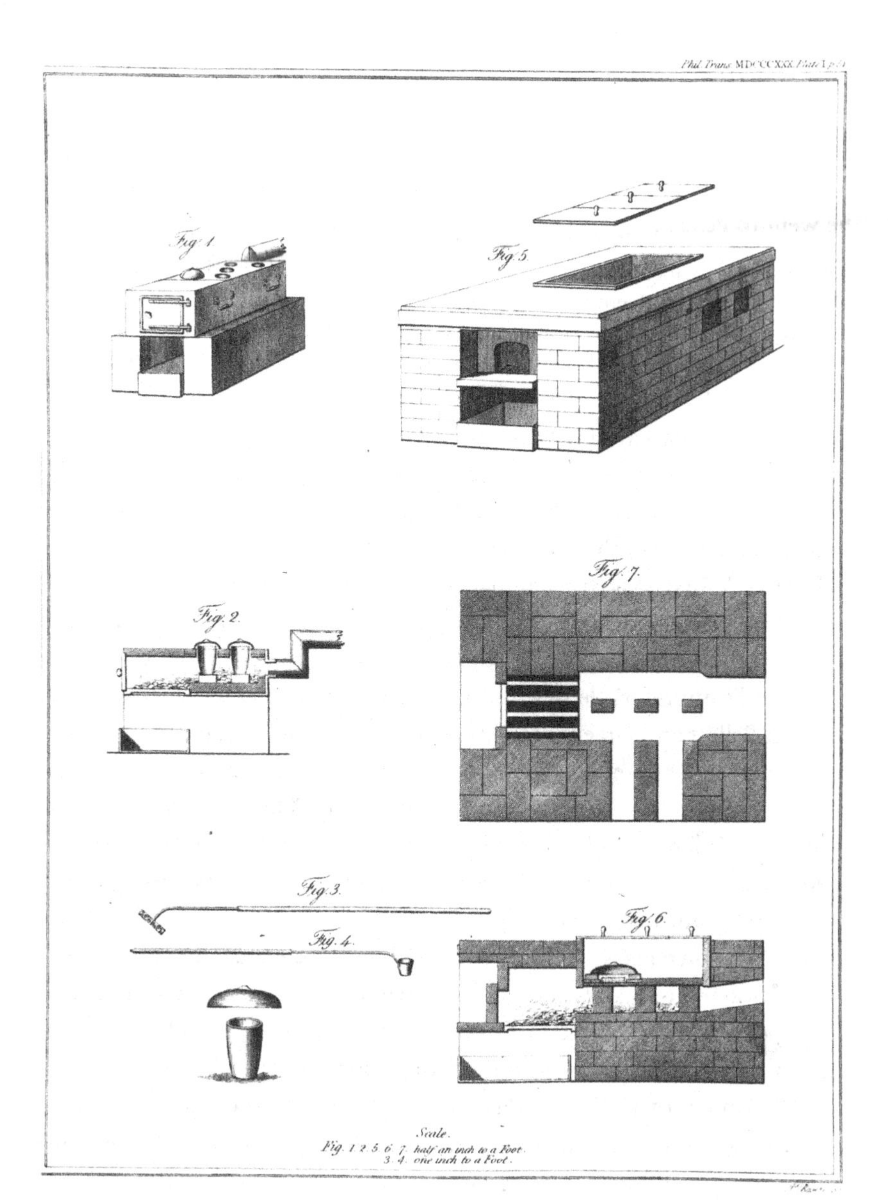

Figure 6.4

Diagram of Michael Faraday's glass furnace at the Royal Institution, 1827. *(Science & Society Picture Library of the Science Museum, London)* Figs. 1 and 2 represent the iron box 30 inches long, 14 inches wide, and 8½ inches deep. The furnace was lined with fire-stone 1½ inches to 2⅝ inches thick. Figs. 3 and 4 depict Faraday's platina stirrer and ladle respectively. Below Fig. 4 is a crucible, which held the molten glass, made of earthenware. Figs. 5–7 represent the finished furnace as it appeared from the front, in cross-section, and from the top, respectively. Surrounded by 18½ inches of brick, it was 64 inches long, 45 inches wide, and 28 inches high. The fire-place, at one end, measured 15 inches long, 13 inches wide, and 11½ inches high. *(Science & Society Picture Library of the Science Museum, London)* merely give suggestions to Pellatt and Green on how to improve their recipe for optical glass, Faraday conducted the experiments himself. In November he went to Pellatt,

> who is perfectly open in his offers of assistance and of unreserved communication on all points [of producing optical glass] with which he is acquainted. He permits me free access to his works at all times that I may gain a knowledge of the general process & particular practises and professes a full readiness to accompany me & explain everything.[38]

One month later, Faraday spent an afternoon and an evening witnessing the operation of emptying, scraping, and charging the glass pots: "My object was to watch the whole of this operation and not to gain any particular knowledge of the materials used. These we know."[39]

Faraday's first task back in 1825 had been the chemical analysis of Fraunhofer's flint-glass lenses procured by the optician Dollond.[40] As Britain's leading chemist during the period, Faraday attempted to reverse engineer Fraunhofer's lenses and prisms by using chemistry; that is, he thought Pellatt and Green, with his chemical analysis in hand, could work backward from Fraunhofer's final products to obtain an experimental procedure that would regularly yield that product. When Faraday received Fraunhofer's lenses and prisms, he immediately proceeded to determine their chemical composition. After ascertaining their ingredients (quartz, potash, red lead, and saltpeter for flint glass; calcium carbonate, sand, and potash for crown glass) and their proportions, he concluded that he could indeed construct equally good glass, since the ingredients were standard and were readily available in Britain. Not surprisingly, Faraday thought that chemistry was the means for creating superior glass blanks.

By December of 1827, Faraday had finally begun experiments to determine the origin, formation, and subsequent diffusion of striae throughout Pellatt and Green's flint glass. At first he ran a series of tests to see if the new glass oven at the Royal Institution was in working order. It was. Faraday altered the ingredients and their ratios in his trials of optical glass. The most promising recipes and the ingredients' proportions were determined by many trial-and-error attempts. In this respect, his work on optical glass resembled Fraunhofer's. Several different recipes were used for each glass batch. The mixture for each recipe, at this stage called the "frit," was kept in a wooden bin. Cullet (pieces of broken waste glass) was added to the frit. This mixture of cullet and frit was heated for up to 24 hours in a coal-burning oven in order to fuse all the ingredients. The mixture was then stirred to ensure homogeneity.[41] The glass blank was then transferred to the annealing oven, allowed to cool, and inspected for bubbles, striae, and transparency. But Faraday was unable to produce transparent, bubble-free, striae-free flint-glass blanks, and his work was also troubled by breakage of the crucibles that contained the glass mixtures, during the intense heatings. Bubbles and striae throughout the glass cause the rays of light to refract through different angles, thereby rendering attempts to produce achromatic lenses futile.[42] The breakage of clay pots during the firing of glass samples, owing to the combination of extreme heat and the caustic nature of the ingredients of glass, caused the ingredients to fuse to the floor and sides of the oven, resulting in a waste of time, labor, and money. After repeated destruction of Stourbridge and Cornish clay pots, Faraday decided to order German Hessian pots.[43]

On 30 August 1828, after 9 months of trying unsuccessfully to produce homogeneous flint glass, Faraday concentrated on Brande's research on the silicated borate of lead for use in optical glass.[44] He commented in his glass notebook: "very dense, hard, & strong, much more so than flint glass. The borate of lead made in this way would form an excellent glass without the addition of any other ingredient."[45] Five weeks later, he was even more confident, looking forward to

> the use of silica & other substances to manufacture glasses in this manageable & convenient way which though heavy shall have such a difference in refractive & dispersive powers as to allow of their being applied in the construction of achromatic glass and so get rid of both ordinary flint and crown glass. Dr. [Thomas] Young says (from the French philosophers) that good flint glass will be of no use unless we manufacture our own good crown glass also.[46]

He found that the most promising recipe for his heavy glass of the silicated borate of lead to be boracic acid (now called boric acid, H_3BO_3), lead oxide, and silica. The lead oxide was originally obtained from litharge (PbO), but this often destroyed the platina[47] trays holding the glass. So Faraday drew upon his chemical expertise and decided to wash the litharge, and dissolve it in diluted nitric acid. This solution was then allowed to cool and crystallize. After 24 hours, the crystals were removed from the liquid solute, examined, broken up, and washed in fresh nitric acid. When the crystals appeared white or bluish white, they were washed three or four times in water to remove any remaining impurities. The crystals were then dried and preserved in glass bottles. The boracic acid was obtained directly from the manufacturer. It had to be white (or bluish white), and entirely soluble in water. The silica used was flint glass makers' sand from the coast of Norfolk, well washed and calcined.[48]

The silica was mixed with the litharge and put into a lidded Hessian crucible, which was then put into the furnace under a bright red heat for up to 24 hours. The glass was allowed to cool, removed from the crucible and pulverized in a Wedgwood mortar. It, too, was washed with water, dried, and placed in bottles. The silica in this state could combine easily with other ingredients to form the glass in the annealing oven. This silicate mixture was then added to lead nitrate and crystallized boracic acid, and was then placed in a crucible in the furnace. When the ingredients melted and fused, the temperature was allowed to rise. A platina stirrer was introduced, and the mixture carefully stirred. The glass was then transferred to an annealing tray and placed in an annealing oven for up to 8 hours. During the annealing, the molten glass was again periodically stirred.[49] The heat was still applied without stirring, and if the glass appeared to have no bubbles or visible striae it was permitted to cool. Good glass samples were then sent to Dollond to be cut and polished into objectives.

Faraday reckoned that the refractive and dispersive ratios of the combined lenses, and not the absolute value of those powers of individual flint and crown specimens, were crucial for optical purposes. Hence, he sought, by introducing fluoric acid (or fluorine) to the borate and the silicated borate of lead mixtures, to create a new type of optical glass whose dispersive and refractive indices were less than those of Fraunhofer's and Guinand's flint glass.[50] Fluoric acid decreases the concentration of lead, thereby dissolving the striae that result from an inadequately dispersed litharge. In short, the borate and the silicated borate of lead increase refractive and dispersive indices of glass, whereas fluoric acid decreases these parameters.[51] Since Faraday's stirring technique was not as successful as Fraunhofer's and Guinand's, he needed to add fluoric acid in order to break up pockets of lead. It was precisely this trial-and-error method that made replication of successful trials so difficult.

Faraday spent 3 years trying to perfect the stirring technique. His approach was rather different from Guinand's and Fraunhofer's. He summarized the problem with the stirring procedure as follows:

> In the glass, the stirring must be in the utmost degree perfect, for if there be the least difference in different parts, it is liable to form striae: nor are the different portions allowed to arrange themselves by their specific gravities, in which case one part might perhaps be removed from another, after the glass was finished and

Figure 6.5
Parallelepipeds of Michael Faraday's heavy glass with four silvered plates, produced at the Royal Institution between 1827 and 1830. *(Science & Society Picture Library of the Science Museum, London)*

cold; but the ascending and descending currents which inevitably take place in the fluid matter, are certain to arrange the irregularities in such a manner as to produce the strongest bad effect.[52]

The analogy he provided the reader was trying to stir a few drops of water added to a clear saturated syrup. Although stirring was necessary to dissolve any waves or striae, it also could be hazardous, if done at the wrong time, for too long, or too briskly, as it would inevitably increase the number of air bubbles forming in the molten glass. Again, Faraday needed to tinker until he found a point at which striae could be eliminated without the introduction of bubbles. Unfortunately, the process often seemed to be hit-or-miss, and Faraday could never come up with a technique that was completely foolproof.

Faraday's stirrer was different from the clay-coated wooden rod of Guinand and Fraunhofer. Faraday used a plate of platina 6¼ inches long and ¾ inch wide for a crucible 7 inches tall. (See Fig. 3 in figure 6.4.) It contained a number of irregularly sized holes, "that, when drawn through the glass like a rake, it may effectively mix the parts."[53] To this platina plate Faraday riveted a thick platina wire approximately 13 inches in length. The end of this wire was screwed into the end of an iron rod, which acted as a handle. The stirrer and the wire were dipped into diluted nitric acid and heated until they glowed red just before being immersed in the glass.[54]

Whereas in Fraunhofer's technique the large stirrer was hooked up to the cover in such a way that the stirring could be carried out with the cover on, Faraday removed the crucible's cover in order to commence stirring. This always ran the risk that a dust particle would fall into the crucible and cause a streak in the glass. The stirrer had to be eased in so that no air would be carried into the molten glass. The chamber was allowed to cool, and the stirring was continued until the glass thickened, at which point the stirring was stopped, the stirrer removed, and the cover replaced. The temperature was then increased for 15–20 minutes, and the process was repeated, sometimes as many as ten times.

Faraday had his assistant Anderson stir the glass manually. This technique required Anderson's hand to come dangerously close to the heat and the molten glass. Fraunhofer's technique only required the clay-covered stirrer to be turned mechanically via rods and pulleys, safely removed from the fire. Faraday had to provide Anderson with a linen bag to put over his hands during the stirring, since "the heat which has to be borne during the operation of stirring, is very considerable, especially on the hands; but at such a moment no retreat from the work, because of mere personal inconvenience, can be allowed."[55]

Bubbles were a much easier problem for Faraday to solve than striae. Once again, Faraday drew upon his impressive chemical expertise:

> It occurred to me also, that . . . [the bubbles'] formation might be hastened and the final separation advanced by mixing some extraneous and insoluble substances with the glass, to act as a nucleus, just as pieces of wood, or paper, or grains of sand, operate when introduced into soda water or sparkling champaign, in which cases they cause the gas, which has a tendency to separate from the fluid, to leave it far more quickly and perfectly than if they had not been present.[56]

At first Faraday used powdered nitre (potassium nitrate) to pull the bubbles down to the bottom of the glass, and Dollond's cut simply removed that portion of the glass.[57] For the silicated borate of lead, however, Faraday employed "spongy platina" for the first time on 13 April 1829.[58] He wrote in his glass notebook:

> introduced some spongy platina, in fine powder & mixed with some rough glass in small fragments. I did this because I concurred that if there was any slow tendency (under the influence of heat) in the glass to give off gas perhaps the powdered platina

acting as nuclei throughout its mass might assist causing the evolution & so more rapidly clear the glass.[59]

Two days later, when the process was complete, he inspected the glass and noted that the samples were free of bubbles. "I have very little doubt," he wrote, "that [the spongy platina] has helped importantly in clearing off bubbles."[60] During the preparation of another batch of glass a month later, Faraday noticed how all of the spongy platina sunk to the bottom (as it should), and that "every piece was surrounded with bubbles. . . . They were adhering to the platina & I think serve to shew its value as a disentangler & separator of them from the glass."[61] Faraday's optimism, however, was to be ephemeral.

Faraday did notice the occasional streak in the finished sample, which he reckoned came from dust particles that had settled in the glass while the cover was removed during the stirring process. Rather than develop a method of stirring that could be carried out while the cover was on, Faraday decided to construct a contraption made of three jars that could act as an air filter. He wrote on 24 October 1829:

I have been trying two or three methods of cleansing the air which passes in at the air tube . . . one plan was by an arrangement of two jars upright one in the other & both with an inverted jar & a tube passed between them . . . so that the air had to pass through a kind of labyrinth. . . . In this way I expected most if not all dust etc. would be stopped within the first or second upright jar.[62]

Striae and bubbles, however, continued to be problematic. Faraday realized during June of 1830 that more stirring was necessary. He wanted to see if he could leave the stirrer in for the entire glassmaking procedure without its reacting with the molten glass.[63] This would allow Faraday to stir more often without the fear of introducing bubbles by the constant insertion and removal of the platina stirrer. Although the stirrer was intact after the procedure was completed, striae and bubbles, which according to Faraday were probably due to dust, were present.[64]

As Faraday produced these glass blanks made of the silicated borate of lead, he sent them to Dollond, who cut and polished them into lenses and passed them on to Herschel, who determined the refractive and dispersive powers of samples by calculating the indices of the extreme violet and red rays.[65] It was Herschel who convinced Faraday just how damaging striae were to optical glass. He sent him a letter detailing, by means of geometrical optics, how striae can affect the dispersive and refractive powers of a glass sample.[66]

In addition to the practical problems Faraday had to overcome, he ran into bureaucratic ones as well. The Board of Longitude, one of the two sponsors of the subcommittee, had been disbanded in 1828. Davies Gilbert had inquired as to whether the Royal Society should continue as the other sponsor of the subcommittee. On 21 July 1828, Herschel had written to Gilbert:

> The Glass Committee . . . is . . . de facto, at an end. It rests with the Royal Society either to let it die then, or to continue as a Scientific Committee of their own body, defraying the Expenses from their own funds, or applying to some other quarter. . . . The present is no doubt a favourable opportunity to terminate the Enquiry if yourself and the Council think it desirable.—My own opinion on that front would be wholly dependent on Mr Faraday's, who is in fact the *bona fide* working member. . . . Mr Dollond has always declared himself ready, if the excise duties were remitted him, to undertake the enquiry as a matter of profitable concern to himself at his own expense.[67]

On the same day he had written to Faraday:

> It strikes me that your opinion as to the probability of success and your *wishes* as to the continuance or discontinuance of the experiments are what will most likely guide the Council of the Royal Society in the question whether to let the subject [of glass manufacturing for optical purposes] drop—or to apply to another quarter for funds to prosecute the enquiry. If success is [to be reckoned] with at all it will be your doing. I am quite sure that I am a non-entity as to all bona fide useful purposes.[68]

And in addition to the dissolution of the Board of Longitude, Herschel's interest in research on optical glass had waned considerably after 1827, culminating in his resigning from the subcommittee in March of 1829 in order to go to the Continent.[69]

Despite these setbacks, Faraday forged ahead with his glass research. Some of the glass samples created by Faraday were cut and polished into lenses and inserted into small telescopes were of very high quality.[70] On 4 December 1828, Faraday wrote in his notebook:

> Called & saw the 4½ inch object glass at Mr Dollond. Understand the astronomers speak well of it.—to me there is a . . . cloudiness not interfering with the goodness of the image rendering it . . . as if a cloud of thin smoke were placed before it. Suppose this less light is not due to reflection to any great degree but to bubbles . . . colour a little . . . both these.[71]

This glass disk, produced during the last week of November 1828, was a result of one of Faraday's first attempts to make use of the silicated borate of lead.[72] Although his technique did not consistently produce good optical glass, Faraday was apparently improving. In April of 1829, glass sample 181, another silicated borate of lead, was sent to Dollond. Faraday wrote: "The piece of glass (181) which I left to be polished has been done & also examined by Mr Dollond. It proves to be the best piece we have made—it is free from bubbles & has only one unimportant stria."[73] Dollond inserted the lens into a small telescope and sent it to George Biddell Airy, then Plumian professor of the Cambridge Observatory. Airy then passed it on to the Italian savant Giovan Battista Amici,[74] who had built an achromatic microscope and who was now interested in astronomical devices.

In 1830 the Royal Society formed the Telescope Committee in response to attacks on English reflecting telescopes being published in the *Astronomische Nachrichten* and in the *Edinburgh Journal of Science* by users of Fraunhofer's refracting telescopes.[75] It was originally hoped that Herschel would test the newly designed British telescopes, which they hoped could rival Fraunhofer's refractors. In March of 1830, Herschel did indeed begin a comparison of results obtained with a British-manufactured telescope with results obtained by Friedrich Georg Wilhelm Struve at Dorpat with Fraunhofer's refractor. Herschel wrote to Gilbert:

> . . . comparisons are odious—especially when they [are] almost of necessity unfair, and tend to no obvious advantage. If Mr. Barlow is in possession of a [liquid-lens] telescope which he is ready to place in competition with Mr. Struve's . . . , I congratulate him on it, and still more the public on his invention of a principle by which such means of observation may be made accessible to men of moderate fortunes.[76]

Herschel concluded his letter with a sentence indicating his total apathy: "*First rate nights* are far too precious to be wasted in trying one telescope against another."[77]

The Telescope Committee reported the tests of refracting telescopes with Faraday's lenses. Astronomer Royal John Pond's 23 December 1830 report on Faraday's lenses was rather encouraging:

> The result has appeared to me extremely satisfactorily, and such as I hope will encourage Mr. Faraday to persevere in his laudable efforts to accomplish the views of the Committee. The Object Glass in question is very achromatic, and indicates that all has been done that could be done by the artist who made the Telescope . . . it shows distinctly many objects usually thought very difficult. The nebula of Orion is shewn beautifully, and I could distinctly perceive the 2 very small stars in the dark oval space. . . . I likewise saw, though with some difficulty, the two smaller stars described by Mr. Peter Barlow, in his late paper which was read at the Royal Society: they lie between the 2 clusters of double stars in σ Orionis. The planet Saturn was not above thirty degrees high when I saw it: the telescope bore as great

a power as can be reasonably expected, but I have not made any elaborate experiments on this point.[78]

Later, however, Pond was to claim that Faraday's telescope was "perhaps deficient in defining power."[79] Captain Henry Kater of the British Admiralty and the Board of Longitude stated in his report of 24 March 1831 that he

viewed Saturn with this telescope; with the whole aperture, the planet was ill defined—but it was much improved by reducing the aperture an inch. On turning the instrument to the nebula in Orion, its great superiority of light over smaller instruments was very striking. Two very small stars which are parallel to the longer diagonal of the Trapezium were distinctly visible & the nebulae was seen very satisfactorily. The small star near Rigel was readily seen, but on one side of the large star was a degree of glaze, evidently arising from some inequality in the construction of the glass. The want of sharpness in the image also must be attributed to fault in the glass as the telescope is perfectly achromatic.[80]

The Royal Society Telescope Committee then requested, in May of 1831, that Faraday "make a perfect piece of glass of the largest size that his apparatus will admit—also that he be requested to teach some person to manufacture the glass for general sale."[81]

Faraday did neither. In a letter addressed to Gilbert written in May of 1830, Faraday clearly stated that he no longer was interested in pursuing his research on optical glass. He wrote: "I . . . wish you most distinctly to understand that I regret I ever allowed myself to be waved as one of the committee. I have had in consequence several years of hard work [and] all the time that I could spare from necessary duties (and which I wished to devote to original research) [has] been consumed in the experiments."[82]

A year later, Faraday echoed these sentiments in a letter to Peter Mark Roget, Herschel's successor as secretary to the Royal Society:

With reference to the request which the Council of the Royal Society have done me the honour of making, namely, that I should continue the investigation, I should under circumstances of perfect freedom assent to it at once. But, obliged as I have been to devote the whole of my spare time to the experiments already described, and consequently, to resign the pursuit of such philosophical enquiries, as suggested themselves to my own mind, I would wish, under present circumstances, to lay the glass aside for a while, that I may enjoy the pleasure of working out my own thoughts on other subjects.

If at a future time the investigation should be renewed, I must beg it to be clearly understood I cannot promise full success. Should I resume it, all that industry and my abilities can effect shall be done: but to perfect a manufacture not being a manufacturer is what I am not bold enough to promise.[83]

Thus, the first British attempt to reverse engineer Fraunhofer's lenses and prisms, with a view to manufacturing them on the commercial level, had come to an end.

The differences between Fraunhofer's and Faraday's glassmaking techniques are informative to historians of science and technology. Because Fraunhofer used wood (as all glassmakers did throughout the German territories at the time), his molten glass was considerably cooler than that of Faraday, who used coal and coke in his glass furnace. As was discussed in the previous chapter, the British had used coal as a source of fuel since the seventeenth century. As a result, British glass of all sorts was produced in major cities, whereas German glass was manufactured in areas where large supplies of wood were present (for example, in the forests of Bavaria and Bohemia). Fraunhofer needed a superior mixing technique that was powerful enough to mix flint glass at a relatively low temperature. He then allowed the glass to set for a period of time, thus permitting the bubbles to float to the top and escape. Another way Fraunhofer decreased the presence of bubbles was to decrease the amount of calcium carbonate added to crown glass.[84] Fraunhofer, drawing upon Guinand's earlier design typical of bellmakers, could stir with the lid on the oven, using a clay-coated piece of wood inserted through the large cover. Also, Fraunhofer stirred his molten glass with a mechanical contraption.

Despite their distinct advantage with the higher temperatures achieved with coal, Pellatt, Green, and Faraday were unable to produce homogeneous flint glass. Faraday's response, as we have seen, was to produce a substitute for flint glass: silicated borate of lead. As this heavy glass was extremely dense, homogeneity problems afflicted him too. Being a chemist and not a glassmaker by training, he had a different set of skills and knowledge than Fraunhofer. Faraday turned to platina for different aspects of his production of heavy glass. Platina was well suited for such research, as it enjoys a very high fusing temperature (over 1600°C) and it does not combine with oxygen, even under the extreme heat required for glassmaking. Because it did not react with any of the ingredients in Faraday's heavy glass, it was used to line the pots and crucibles, which stored the glass while in the furnace, as well as the ladle used to transfer the molten glass. Furthermore, spongy platina was used to combine with the bubbles that formed during the stirring and to draw them to the bottom of the pot. And platina was used as the stirring rod for the molten glass. Faraday's stirring technique must not have been as good as Fraunhofer's, as the largest lens he produced was 4½ inches in diameter. The diameter of Fraunhofer's Dorpat refractor was 9⅗ inches. And recall that during the last year of his life Fraunhofer was working on a 12-inch object lens for an equatorial. Also, Faraday experienced homogeneity problems with pots holding only 8 or 9 pounds of glass, whereas Fraunhofer's batches were between 400 and 500 pounds.[85]

Of course, Fraunhofer was an optician and a glassmaker; Faraday was neither. Optical glass was Fraunhofer's livelihood, whereas it distracted Faraday from his personal research interests (which certainly did not include manufacturing optical glass). Fraunhofer possessed all the requisite skills to make achromatic lenses. He produced the crown and flint glass. He cut, ground, and polished lenses from glass blanks. He determined the refractive and dispersive indices (more accurately than anyone else during that period) for each colored ray for each glass sample. The division of work in Britain mirrored the labor aristocracy of London's glassmakers. Faraday executed the physical production of the glass, while Dollond cut, ground and polished Faraday's promising samples into lenses for telescopes. Herschel calculated the dispersive and refractive powers of the glass samples, and offered his advice to Faraday concerning the geometrical optics relevant to homogeneity.

Liquid Lenses

During the late 1820s, different techniques of manufacturing achromatic telescopes emerged from attempts to circumvent the British inability to produce Fraunhoferian-quality optical glass. For example, in 1828, Alexander Rogers of Leith proposed the construction of a refractor whose achromaticity depended neither on the superior quality of large flint-glass lenses nor on a very precise knowledge of the refraction and dispersion of the kinds of glass used, but on a clever geometric combination of several lenses. The telescope was constructed by interposing between a single object glass and its focus a solid compound lens, free from refraction by the opposing powers of a convex lens of plate glass and a concave lens of flint glass, but possessing a dispersion equal to the difference of the dispersions of its component lenses.[86] Rogers's suggestion was never pursued. But there already existed a technology that did seem promising to some. Since the late eighteenth century, liquid lenses had seemed to be the way forward. During the 1820s, two individuals began to specialize in the production of achromatic telescopes using liquid lenses. Liquid lenses had a marked advantage over flint-crown doublets in that instrument makers could shorten the length of a telescope by increasing the dispersion of light, since the refractive index of the fluid is nearly the same as that of the best flint glass, whereas its dispersive power is more than twice that of flint.[87] That is to say, since the dispersive power of the fluid was very great, one could place the two objectives at a considerable distance from each other. And the second lens with the sulfuret of carbon (now called carbon sulfide), which was normally the flint glass in a refracting telescope, could be much smaller than the first.

Archibald Blair, whose father, Robert Blair, had developed liquid lenses for telescopes in the 1790s, continued his father's work; according to Brewster's notice in the *Edinburgh Journal of Science* in 1827, Archibald's telescopes were vastly superior to all ordinary achromatic instruments.[88] Peter Barlow of the Royal Military Academy in Woolwich also attempted to avoid using flint glass in achromatic telescopes by using liquid lenses.[89] His smallest telescope, with a 3¼-inch aperture, could distinguish all the double stars of the class that Sir William Herschel listed as tests for 3½-inch achromatic telescopes.[90] Brewster's notice in the *Edinburgh Journal of Science* concluded with the hope that Barlow would "submit these instruments to the Board of Longitude, whose special duty it either is, or ought to be, to patronize with a liberal and active zeal every improvement on the telescope." Brewster continued: "If other nations have already been allowed to outstrip ours in this branch of rival manufacture, the time has come for retrieving our character, and replacing us in the position from which we have been driven."[91]

Liquid lenses, Barlow argued, could in theory overcome chromatic aberration because an appropriate fluid could be mixed to refract the variously colored rays of the spectrum in the same proportion as flint glass. By 1827 a fluid had been found, namely sulfuret of carbon, which could correct chromatic aberration by dispersing the colored rays through the same angle as flint glass did, and hence could be coupled with a crown-glass lens in order to correct achromatic aberration.[92] Although liquid lenses might now seem to be far-fetched, and perhaps ill conceived, they did possess some theoretical advantages, as Barlow argued using the results obtained with his fluid-based telescopes in 1827.[93] In 1828 he claimed that the sulfuret of carbon had every requisite he required.[94] Most important, in contrast to Robert Blair's liquid-lens telescopes, in Barlow's the liquid did not need to come into immediate contact with the two or three lenses composing the objective. Because the dispersive index of the sulfuret of carbon was so great, Barlow could place the fluid correcting lens, which had to be only half the diameter of the crown glass, at a distance from the crown glass of half its focal length. This distance would guarantee achromaticity at higher powers of magnification. And since the diameter of the fluid lens was reduced, the focal power of the instrument increased at least 1.5 times; hence, the telescope could be reduced to two-thirds the length of a telescope with conventional glass lenses without incurring a greater amount of spherical aberration in the front lens.[95] As flint glass was to be avoided since it was so difficult for the glassmaker to produce and its lower dispersive power required larger telescopes, liquid-lens telescopes seemed, at the time, a powerful alternative to Fraunhofer's refractors.

Barlow recalled how he "saw so much the difficulty which opticians experience in obtaining large pieces of good flint-glass, that [he] turned [his] attention to supplying this material by a fluid."[96] As was general procedure at the time, Barlow used William Herschel and James South's catalogue of stars in order to test the precision of his instrument. He proudly announced that η Persei, which had been catalogued as a double star with a small star at a greater distance, could be seen distinctly as a sextuple in his telescope. He added: "These stars I had the satisfaction of showing to M. Struve in his recent visit to England."[97] Also, σ Orionis, which had been catalogued as two distinct sets of stars, each triple, was seen as two sets of quadruples with two very fine stars between them.[98]

In 1828, Barlow requested funds from the Council of the Royal Society to construct a giant achromatic telescope using sulfuret of carbon liquid lenses. After receiving a letter from Barlow early in the year, the board agreed to advance him £200 toward the expense of completing "his invention of an improved achromatic telescope to be made without flint glass."[99] The council agreed and ordered Dollond to construct a telescope under Barlow's supervision.[100] Barlow continued his research. In 1833, Herschel, G.B. Airy, and Captain W. H. Smyth all provided detailed accounts of Barlow's telescope. It was generally reckoned that it still suffered from chromatic aberration under high magnification, and that, contrary to Barlow's claims, more work was needed.[101]

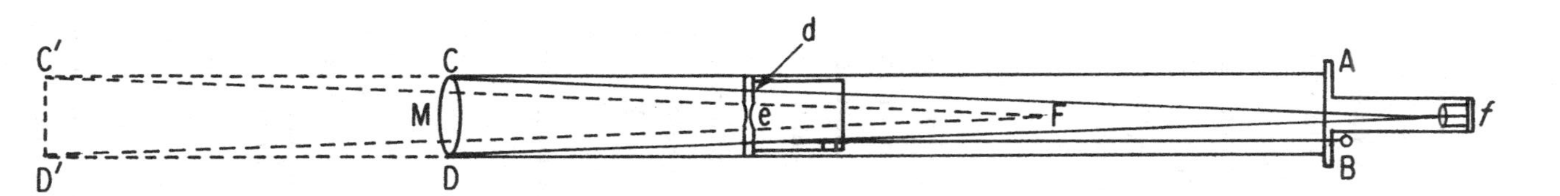

Figure 6.6
Peter Barlow's liquid-lens telescope. ABCD represent the tube of a 6-inch telescope. CD is the object glass; F is the first focus of the rays, and de the fluid concave lens, which is 24 inches away from CD. The focal length MF is 48 inches, and the diameter of the fluid lens is 3 inches (48 is to 6 as 24 is to 3). The resulting compound focus is 62½ inches. The rays df and ef converge at the focus with the same intensity of light as if they proceeded from a lens with a 6-inch diameter, placed beyond the object glass CD determined by tracing the rays until they touch the sides of the tube at C'D', or 62½ inches beyond the fluid lens. The tube is thereby shortened by 38½ inches. The length of the refractor is shortened nearly half.

Hence, it became clear to the British through the course of various investigations by several experimental natural philosophers that the reproduction of Bavarian-quality flint glass was not possible. One needed more than a recipe to construct an achromatic lens; certain skills were also necessary. A culture responds to the needs of society by utilizing its own resources and technologies. British attempts to construct achromatic lenses centered on the total circumvention of the use of flint glass, either by using Faraday's heavy glass composed of the silicated borate of lead in combination with crown glass or by drawing upon Barlow's and A. Blair's revival of the technology of liquid lenses in order to create an achromatic telescope.

Secrets and Bribes

In addition to Faraday's invention of a new glass technology and the work on liquid lenses, the third attempt to improve achromatic-lens production in Britain involved purchasing secret information and even resorting to bribery. Back on 29 September 1821, as was briefly mentioned in the previous chapter, M. Reynier, Pierre Louis Guinand's assistant, sent a letter informing the Council of the Astronomical Society of London that he was "in possession of a process for making disks of flint glass, fit to be employed in the construction of object glasses for achromatic telescopes."[102] Reynier was invited by the Council to send some of those flint-glass disks, which he did. The first set was "very unsatisfactorily" small (only 2 inches in diameter), and therefore it could not be properly adjudicated. The Council, it turns out, was more interested in the procedure of manufacture than in the products:

> The inquiries of the Council too were answered in a manner hardly more satisfactorily, M. Guinand appearing chiefly desirous of disposing his stock of discs on hand, and that a very limited one, at a tarif annexed, and of obtaining a pecuniary compensation for his secret, rather than of continuing the manufacture.[103]

Guinand and Reynier then sent a flint-glass disk 7½ inches in diameter, stunning the Council, which had not seen a flint-glass objective lens with a diameter greater than 4½ inches. On 14 November 1823 the disk was handed over to Dollond and Charles Tulley. Tulley was given the task of producing a concave lens of crown glass to be combined with Guinand's lens of flint glass. It took Tulley more than 2 years to produce a piece of crown glass homogeneous enough to match Guinand's flint glass. Once completed, the

Figure 6.7
Peter Barlow's telescope with a liquid lens, built in Woolwich. It possessed an aperture of 7.8 inches, an inch greater than that of the largest refractor in Britain at the time. Its tube was 11 feet long, and its eyepiece added a foot. Owing to the liquid lens, its effective focus was 18 feet. The stand weighed 400 pounds, the telescope itself 130. See Barlow 1829, p. 33. *(Syndics of Cambridge University Library)*

telescope proved "in the highest degree satisfactory." It even passed the litmus test of the period for achromaticity:

> The examination of a bright object on a dark ground, as a card by daylight, or Jupiter by night, which high magnification powers affords as is well known, the severest test of the perfect achromaticity of a telescope, by the production of green and purple borders about their edges in the contrary case. The telescope in question bears these tests remarkably well, and is certainly more achromatic than usual, a circumstance depending not merely on the nice adjustment of the foci, but on the quality of the flint glass.[104]

At a meeting of the Council on 17 March 1823, Gilbert, Herschel, and William Pearson decided not to pursue the matter any further, as there was no guarantee of a regular supply of flint glass from Guinand. As it turns out, Guinand was gravely ill; he died less than a year later.

Four years later, Guinand's son, Henri, wrote to Herschel offering to travel to London in order to show Herschel some of the glass samples he had produced from his deceased father's glass recipe. Herschel replied:

> You should know that I am myself at present acting as a member of a scientific committee for the *discovery & publication* of a process for making glass exempt from veins. It is my duty therefore to caution you [—] should you think proper to correspond further with me—against making me by inadvertence the depositancy of any kind of the details of a secret which you consider so valuable and which should they become known to me, I shall be bound to make public.
>
> I have no connexion with any manufacture of glass further than owing to the circumstances I have mentioned.[105]

This passage nicely illustrates Herschel's moral principles. As an experimental natural philosopher, Herschel felt morally obligated to disclose any knowledge that would benefit the scientific community. However, he also felt obliged to inform Henri that he would indeed pass on any relevant information.

Henri Guinand never traveled to London, but he wrote to Tulley requesting 6000 francs for the secret of producing optical glass. Tulley wrote to Herschel that the fee was too much for an optician alone to pay but that the Astronomical Society, the Royal Society, or the Board of Longitude might consider paying the fee "for the benefit of science generally."[106] Tulley continued by warning Herschel not to pay Guinand the money until Henri revealed his experimental results and shared his secret of manufacturing, but also stated that he did "not think his [Henri's] terms too high for the secret."[107]

The British had already resorted to desperate measures, including bribery, in 1825, when Fraunhofer had been offered as much as £25,000 for his secrets.[108] Fraunhofer did not take the British up on their offer, as he held to his written pledge to Utzschneider that any knowledge related to the production of achromatic glass was the Optical Institute's property and therefore was not to be revealed to anyone other than the institute's apprentices (and indeed then only sparingly). The Joint Committee had then turned to Guinand's family members. In a meeting on 24 May 1827, just after Herschel had written to Gilbert concerning the persistent failure of Pellatt and Green to make flint glass, and two weeks after recommending that the Royal Institution take over the operation, it was "resolved that previous to incurring further expense in the manufacture of flint glass for optical purpose, the Board of Longitude be applied to, to authorize a personal application to the relatives of the late Mr Guinand, with a view to obtaining that portion of information which they may possess beyond ordinary glass manufacturers."[109] After that attempt also failed, the Royal Society committee decided to try to obtain information from the French optician Georges Bontemps. Bontemps had previously paid Henri Guinand a significant sum of money for his father's optical glassmaking formula. The minutes of the 13 November 1828 meeting read: "resolved that Capt. [Henry] Kater be requested to make inquiries respecting the terms on which a knowledge can be obtained of the secret method employed by the late [*sic*] M. Bontemps[110] of making glass for optical purposes."[111] From 8 December 1828 until 6 January 1829, Kater, Roget, and Bontemps corresponded.[112] On 8 December 1828, Bontemps wrote to Kater discussing his and Mr. Chibaudeau's ability to manufacture optical glass for telescopes of large dimensions. According to Bontemps, his glass could rival the late Guinand's in terms of transparency and homogeneity.[113] Kater immediately responded by penning Gilbert that he forwarded proposals from Bontemps "for the sale of the secret for making glass for optical purposes. . . . I beg to add that M. Bontemps has not received from me the slightest encouragement to expect the purchase either of the secret or the Dishes [the glass specimens that Kater purchased for examination by the committee], but rather otherwise."[114] Bontemps's samples were, much to the outrage of the committee, of grossly inferior quality. Dollond analyzed the glass and reported to Roget that "the largest disk is extremely bad, and in my opinion very improper for the object-glass of a telescope. The smallest disk is not quite so imperfect as the largest, but in my opinion of little value."[115]

By late 1827 the English were beginning to realize that their attempt at reverse engineering was a failure. Despite Faraday's limited success using the silicated borate of lead from 1828 to 1830, he never produced lenses the size of either Guinand's or Fraunhofer's. Since the members of the subcommittee believed that they knew all the ingredients of Guinand's and Fraunhofer's glass from Faraday's chemical analysis, they began to wonder whether the actual practices and labor of glass manufacturing might not be recoverable from the lenses themselves. They also turned to Fraunhofer and Guinand themselves, as well as Henri Guinand and Bontemps for assistance. As was discussed in the previous chapter, whether or not skilled artisans' practices of achromatic-lens production could be communicated, and if so how, became a central concern of British experimental natural philosophers during the 1820s and the 1830s. Certainly Faraday and Herschel believed that they could produce glass and lenses at least as good as Fraunhofer's. Indeed, Faraday's first task for the subcommittee, the chemical analysis of Guinand's lenses (and later Fraunhofer's), is a powerful indication that he believed that chemistry underpinned glass production. Interestingly, he originally thought that the recipe's ingredients were key to success, implying that the actual practices involved in the production of glass were subordinate issues. As we have seen, this view was highly contentious, and Faraday himself reversed his stance in 1830, after several years of failure, by questioning whether Fraunhofer's and Guinand's knowledge was indeed practical, private, and not of a nature to be communicated.

7

The End of Fraunhoferian Hegemony

And with him [the skill of genius] dies, awaiting the day when nature once again endows another in the same way. . . . [The rule of art] cannot be one set down in a formula.

—Immanuel Kant, *The Critique of Judgement*

Mühlbeck [the village glassmaker] is dead, no one knows the secret of ruby glass.

—The Old Man, in Werner Herzog's *Herz aus Glas*

During the first week of May 1826, Joseph von Fraunhofer lay on his deathbed. During an early-autumn journey down the Isar River from Tölz to Munich, he had contracted a lung disease. Such ailments commonly afflicted glassmakers, as prolonged exposure to both the intense heat of the furnaces and lead oxide—which destroys the lungs—was detrimental to their health. Utzschneider was distressed. The future of his Optical Institute looked bleak indeed. Fraunhofer, like all glassmakers of the early nineteenth century, never wrote down formal instructions for the production of optical glass. During that first week in May, Utzschneider ordered Fraunhofer's close friend, the royal director of the mint Heinrich J. von Leprieur, to sit beside Fraunhofer's bedside and record the glassmaking procedures and recipes from the gravely ill glassmaker.[1]

These notes taken by Leprieur on Fraunhofer's secrets have an astonishing history. On 7 June 1826, the day after Fraunhofer's death, they were removed "in a special manner [i.e., under the watchful eye of Leprieur] from Fraunhofer's apartment."[2] On 27 September these papers were handed over to the Ministry of the Interior for safekeeping. Only Utzschneider was permitted access to them. On 19 December, Utzschneider, having received permission from King Ludwig I, requested that the Ministry of the Interior seal the papers and restrict access to them.[3] During the winter of 1838–39, the institute was taken over by Georg Merz and Josef Mahler, Fraunhofer's apprentices. Although Fraunhofer's papers were to become their property and responsibility, Utzschneider died before completing the arrangements. Merz and Mahler then petitioned the king to ensure that Fraunhofer's papers and all their own records would be stored in the Bayerische Hausarchiv, "so that secret of Fraunhofer's preparation of crown and flint glass could not be taken up by foreigners."[4] Only the successors of Utzschneider's and Fraunhofer's Optical Institute were permitted access to Fraunhofer's secrets. In 1843, after Mahler's death, the minister of state wrote to both the minister of the interior and the foreign minister to inquire whether the Hausarchiv should continue to guard the transcripts, "in order to safeguard the discovery of the Fatherland [i.e., Bavaria]." The Bavarian minister of state realized that "on the one hand [keeping the transcripts secret] would ensure the continuation of the Optical Institute's hegemony and make it more difficult for another country to take over, but on the other hand [there was the distinct chance that] the method of production would fall into oblivion." It was decided that Fraunhofer's transcripts should now become the property of the state. Fraunhofer's secrets were too important for dissemination, even 17 years after his death. The Bavarian state ruled that the recipes and the manufacturing techniques for optical glass would remain in Merz's custody if he agreed not to leave Bavaria, in which case the archives would be returned to the Hausarchiv in Munich. Fraunhofer's secrets stayed with Merz until 1918. After the end of World War I, the Georg and Sigmund Merz Optical Company, which had moved to Pasing, declared that Fraunhofer's transcripts no longer possessed any scientific value and that access to them was no longer dangerous.[5] In 1925 (99 years after Fraunhofer's death), the chemical and technical department of the Bavarian Institute of Trade (Landesgewerbeanstalt) concurred. In a report on "the present-day practical utility of Fraunhofer's papers," the committee argued that the transcripts "do not possess any entrepreneurial relevance to those individuals and scholars interested in modern glass manufacturing."[6]

The Beginning of the End

After Fraunhofer finally succumbed on 6 June 1826, Utzschneider went to great lengths to proclaim that, despite the death of the world's greatest optician, his Optical Institute would continue to provide lenses of superior quality to its anxious clients. In the *Augsburger Allgemeine Zeitung*, Utzschneider attempted to allay his patrons' fears:

> If a few of you believe that the spirit of the Optical Institute has been buried with Fraunhofer, you are gravely mistaken. Fraunhofer's spirit is still alive and is in full activity. Even though the Privy Councilor von Utzschneider is already sixty-three years old, he still possesses youthful vigor. And he will not hesitate—as a result of his generally well-known patriotism, love of art and science, and knowledge of the secret of flint and crown glass—to hand over this art of flint and crown production to a young Bavarian [Georg Merz] so that this knowledge will never be lost again."[7]

After deciding against hiring either of the instrument makers Carl August Steinheil and Friedrich August Pauli to be the new director in charge of optical glass production,[8] Utzschneider appointed Merz, Fraunhofer's apprentice, who had worked for his master for 18 years.[9] Before Utzschneider gave Merz the official offer, however, he gave him a quiz consisting of six problems: to list the properties of both crown and flint glass, to explain the determination of the relationship between the refractive and dispersive indices, to explain the construction of a lens system that will correct spherical and chromatic aberration, to explain the practice of lens making in relation to optical theory, to explain the classification of different types of glass, the experimentation (Untersuchungen) of the various types, and their assembly into a lens system, and to explain the centering of glass lenses so that they share a common axis.[10] Merz responded quickly and to Utzschneider's satisfaction and was subsequently given the directorship of achromatic glass production at the Optical Institute.[11]

Utzschneider used the notes taken by Leprieur at Fraunhofer's deathbed to carry out his experiments with optical glass, presumably alongside Merz. There are detailed protocols from two trials, one dated 19–24 February 1827 and one dated 19 March 1829.[12] From the February 1827 trial, we see that Utzschneider was having some problems with glass homogeneity and color. But as time went on, and with Merz at his side, the glass samples steadily improved in quality.[13]

Utzschneider had hoped that Merz would be well suited to continue the institute's legacy, after learning from the master for so many years. Merz had grown up in the surrounding community of Benediktbeuern and was taught mathematics and astronomy by the Benedictine monk Rauch.[14] He was able to continue the work that Fraunhofer had started and to carry out work on new orders. But after two decades, international orders began to shrink considerably. Hence, Merz enjoyed limited success. His apprenticeship with Fraunhofer enabled him to proceed after Fraunhofer's death, yet apparently he was not as gifted as his master. From 1826 to 1843 the Optical Institute received orders for twenty large refractors from locations around the world.[15] Included in those orders were Merz and Mahler's greatest accomplishments: the object lenses for both the Pulkovo refractor (built in 1840) and its twin (completed 3 years later for Harvard College). But those two refractors, the largest in the world at the time, were their only claims to fame. Both men were accomplished opticians and Mechaniker, having learned their trade under Fraunhofer, but their skills were not sufficient to enable them to produce glass of the quality that had brought their Meister world renown.[16]

A loss in credibility resulting in a steady decline of orders inevitably led to a decrease in the quality of manufacture. After the Optical Institute in Munich made the Pulkovo and Harvard refractors, its reputation declined rapidly. This was painfully evident at the Great Exhibitions of the 1850s and the 1860s. Merz and Sons failed to win the Council Medal for the best specimen of optical glass at the Great Exhibition of London in 1851.[17] The jury reported that "although the workmanship of those [astronomical instruments] exhibited by Germany deserves the highest prize; the instruments are . . . few in number, and do not fully represent German art."[18] They had seen better from the Germany territories before.

All in the Family

Fraunhofer's legacy was not limited to the Optical Institute in Munich. Recall that after leaving Benediktbeuern as a result of a personality conflict with Fraunhofer in late December of 1813, Pierre Louis Guinand returned to his hometown of Les Brenets in Neuchâtel, Switzerland. Two years later he bought a glass hut there and began manufacturing achromatic lenses with wife, Rosalie, and his elder son, Aimé. Both Rosalie and Aimé had assisted Guinand and Fraunhofer in achromatic glass production back in Benediktbeuern, and had manufactured achromatic lenses good enough to catch the attention of Tulley and Herschel in 1821. Aimé continued to manufacture glass after his father's death.[19] Aimé's son-in-law, Théodore Daguet, worked with both Rosalie and Aimé at Les Brenets. He set up his own glass hut at Solothurn, which produced some of the world's finest optical glass.

Meanwhile, Aimé's younger brother, Henri, decided that his father's business could provide a lucrative future. After his father's death, he attempted, unsuccessfully, to manufacture blanks of achromatic glass using his father's recipes. But his failure did not abate his desire for financial reward. He proceeded to sell his father's secrets in 1827 to the opticians Thibeaudeau, Noël Jean Lerebours, and Georges Bontemps of Choisy-le-Roi, France.[20] Of particular importance was the secret of the skills involved in the stirring process that Guinand had developed and Fraunhofer had perfected at Benediktbeuern. Henri offered his services, which included the secret, for 3000 francs. After originally balking at the price, Thibeaudeau and Bontemps agreed that Henri should be in charge of the technical production of the flint glass from March 1827 until March 1828. After one year, Henri failed to replicate his father and Fraunhofer's glass, and the contract was thus terminated.[21] Bontemps successfully manufactured optical glass disks at Choisy-le-Roi in the autumn of 1828, and they were put on display in the Académie des Sciences.[22] Optical lenses were manufactured at Choisy-le-Roi from that time on. In 1836, Henri produced a disk of flint glass 14 inches in diameter. Bontemps manufactured a crown glass complement for it in 1843. The objective lens constructed by Lerebours in 1849 from these glass specimens became the property of the Observatoire de Paris.[23]

The French opticians' reputation soared in proportion to the decline of the reputation of Utzschneider's Optical Institute. In 1829, Utzschneider wrote a letter that was published by Schumacher in his *Astronomische Nachrichten* and subsequently translated and published in English. Utzschneider wished, first and foremost, to dispel "some unfavourable impressions which have got abroad relative to my manufactory of glass for optical purposes."[24] He began his letter, which was addressed to Schumacher himself, by denying that Thibeaudeau and Bontemps had "rediscovered the secret of producing flint glass of any magnitude, highly favourable for optical purposes; a secret which they pretend has been lost since the death of Fraunhofer, and Guinand the father: and that amongst the pieces presented to the Academy of Sciences [in Paris], there were some of fourteen inches in diameter."[25] Utzschneider also denied that Pierre Louis Guinand was responsible for the production of optical-quality flint and crown glass. Utzschneider stated that before he had hired Guinand at

Benediktbeuern in 1806 he had forced the Swiss bellmaker to divulge all his secrets concerning optical glass production and

was convinced that his efforts would not have been attended with any advantage either to science, or to his own interests. . . . The description of the castings of M. Guinand . . . proves that he would not have succeeded but for the experiments made with me at Benedictbeurn, and at my expense. Still the glass of the last casting, which was made at the commencement of the year 1814, was not equal in quality to that which Fraunhofer made at a later period.

The flint glass for the object glass of the Dorpat telescope was not cast till four years after the departure of M. Guinand. . . .

In a letter of February 10, 1816 he [Guinand] offered me his services [after having departed just over two years earlier from Benediktbeuern], stating that "I have recently put it in practice by two small castings." But, M. Guinand, at that time, still did not know how to produce glass for optical purposes. . . . Since [Fraunhofer's] death, I have myself undertaken the continuation of the manufactory of glass destined for optical purposes; and I believe that I can guarantee its excellence. The object-glasses recently constructed by my workmen sufficiently well attest that the secret of making flint glass, of any size, for optical purposes, is not yet lost, as the Globe [a leading newspaper of the period] would have us believe.[26]

Utzschneider, quite clearly, wished to preserve historically the achievements of his friend Fraunhofer. More important, he also wanted to ensure that his Optical Institute would still receive orders despite Fraunhofer's death. Just how much Pierre Louis Guinand's work on glass contributed to Fraunhofer's and the Optical Institute's success seems impossible to recover. Precisely this question was raised at the time. In 1825 Herschel wrote to Schumacher:

a great sensation among our opticians & telescope fancyers has been produced by a little work of Mr Reynier, translated by Col. de Bosset [:] a biographical sketch of M. Guignand [*sic*] the glass-manufacturer of Neufchatel. He claims the invention of Frauenhofer's [*sic*] process, and if the facts there stated be true, has certainly a priority in his favour, though F[raunhofer] has probably *since* made great *improvements*.[27]

Historically, however, it is interesting to note that the claims as to who possessed the necessary secrets of producing optical glass were certainly nationalistic in origin. French accounts, by Bontemps for example, stressed Guinand's role, while German stories, such as those by Helmholtz and Abbe (discussed in the next chapter) and Utzschneider's, credited Fraunhofer with the artist-genius ability to make superior artifacts. Moreover, such accounts served the commercial advantage of those advancing them.

After his failure with Bontemps and Thibeaudeau, Henri Guinand entered into an agreement in 1848 with his grandson, Charles Feil, a Parisian optician.[28] This partnership eventually led to France's premier optical enterprise: that of the Feil Company, led by Charles and his son Edmond. The Feil Company was to become Carl Zeiss's main competitor. In an attempt to procure some form of reward for his late father's labors, on 25 June 1838 Henri divulged the secret method of glass stirring to two members of the Paris Academy, D. F. J. Arago and J. B. Dumas.[29] Eventually, Henri received the award he had so desperately sought. On 13 August 1838 he was awarded the Lalande Medal for the communication of his father's and Fraunhofer's secret. In addition, he shared the Paris Society for the Advancement of Industry's annual prize of 10,000 francs with Bontemps, who had clearly improved his method for the production of flint glass over a ten-year period. These men also shared the 8000-franc prize for the best piece of crown glass, also sponsored by French industry.

Fraunhofer and Guinand's legacy also made its way to England. Bontemps was obliged to leave France owing to circumstances connected with the Revolution of 1848. At the time, no one in the world knew more about producing optical glass than Bontemps. In April of 1848 he arrived in Birmingham at Chance Brothers, Britain's leading glass firm, where he agreed to superintend the colored and ornamental glass departments, to direct the manufacture of optical glass in accordance with James Chance's patent of 1838 for the polishing of sheet glass and to "advise generally and assist in the glass business" of the firm. Bontemps received an annual salary of £500, plus five-twelfths of the net profits of the optical department and one-tenth of those of the ornamental department.[30] Bontemps had been well known to the Birmingham firm since his 1828 visit in connection with his method of producing large achromatic lenses for telescopes, which he had just obtained from Henri Guinand.[31]

Fraunhofer and Guinand's French and British successors were far more successful in optical-lens production than their Bavarian direct descendants, Merz and Mahler. It is noteworthy that when the Chance Brothers of Birmingham wanted to improve all aspects of glass production, in the 1830s, they turned to the French glassmakers and not the Germans. After his tour of glassmaking firms in Prussia, Bavaria, and Bohemia, John Reynell of the Chance Brothers firm concluded in 1833 that French glassmaking was superior to German.[32] Skilled labor was of paramount importance for the

English firm, since the excise duty financially prohibited experiments with crown and flint glass. Although French glassmakers possessed the necessary skills, problems often arose when they worked in England. Lucas Chance found that the French not only demanded higher wages than the British but also could be insubordinate. Their unwillingness to live in a foreign country and learn a foreign language, and their guild customs, which prohibited instructing outsiders in the art of glassmaking, convinced Chance that the English should train their own labor force.[33]

The Chance Brothers' employment of Bontemps had guaranteed their monopoly in the British market. It was, after all, the Chance Brothers firm that received the order for the glass of Joseph Paxton's Crystal Palace in 1851.[34] It was at the first Great Exhibition that the Chance Brothers attempted to enter the Continental market in competition with French firms. They received the highly coveted Council Medal for a perfectly manufactured 29-inch disk of flint glass weighing roughly 200 pounds. As the jury reported, Merz and Mahler's "great object-glasses of Pulkovo and of New Cambridge [the Harvard College Refractor] did not exceed 16 inches."[35] In the lenses and prisms competition, the French fared better than the British, however. A Council Medal was also awarded to Daguet for a large sample of flint glass. The only Council Medal awarded for a telescope was given to Buron of France. Fraunhofer's successor, Merz, received only one Council Medal, for an equatorial.[36] The monopoly that the Munich Optical Institute had previously enjoyed under Fraunhofer's direction was now lost. Bontemps was indeed struck by how quickly the German territories fell from sight in the optical race. During his Continental journey of 1850, Bontemps remarked how German opticians lacked ambition, and he hoped to introduce the Chance Brothers' optical glass to them.[37] Ironically, most opticians during the 1840s and the 1850s were using flint glass produced by Aimé Guinand's son-in-law, Daguet. Hence, Bontemps was pleased to hear from telescope maker A. Ross that his flint glass was "even superior to the Swiss flint."[38] Bontemps was able to procure £200 worth of orders from German opticians.[39]

At the Paris Exhibition of 1867, Chance Brothers had the only British exhibit of window glass. Their newly developed lighthouse apparatus was a success. Their optical exhibit was second only to that of Feil. Two years later the Chance Brothers firm was elected to the French Académie Nationale Agricole, Manufacturière et Commerciale and awarded the

Médaille d'Honneur.[40] By 1860, Daguet's glassmaking firm in Solothurn had stopped producing optical lenses; hence, there remained two major suppliers of achromatic lenses to the world's scientific communities: Chance Brothers Company of Birmingham and the Feil Company of Paris. Although it is rather difficult historically to pinpoint the exact cause of the success of the Chance Brothers Company, clearly Guinand and Fraunhofer's secrets (passed on by Bontemps), the repeal of the excise tax, and a very successful managerial policy all contributed to the company's fame and fortune.

In short, from about the early 1830s until the early 1880s, optical hegemony had left German soil. Hermann von Helmholtz and Ernst Abbe, we will see, pleaded with the Prussian government to subsidize Abbe and Otto Schott's enterprise at the Carl Zeiss Works in Jena, using Fraunhofer as a historical example to argue that Germany had once been the home of the world's greatest optical institute. But after Fraunhofer's death, rival nations now manufactured prized optical glass. As will be discussed in the next chapter, Helmholtz and Abbe realized that a united Germany could now achieve what reformers had attempted more than 50 years earlier: science could unite the new, culturally diverse nation.

8

Forging an Artisanal History and a Cultural Icon

Ehret Eure deutschen Meister:
Dann bannt ihre gute Geister!
Und gebt ihrem Wirken Gunst;
Zerging in Dunst
Das Heil'ge Römische Reich,
Uns bliebe gleich
Die heilige deutsche Kunst.

[Honor your German Masters
Then keep their good spirits alive
And cherish their deeds.
The Holy Roman Empire
Vanished into thin air
For us still remains the Holy German Art.]

—Richard Wagner, *Meistersinger von Nürnberg*

On the eve of 6 March 1887, in the hall of the Berlin Rathaus, a packed audience that included some of Germany's leading scientific and political personalities was treated to an elaborate celebration of the hundredth anniversary of Fraunhofer's birth. The audience included Leopold Loewenherz (editor of *Zeitschrift für Instrumentenkunde*, soon to become the first director of the technical section of the Imperial Institute of Physics and Technology [Physikalisch-Technische Reichsanstalt]), Rudolf Fuess (technologist, entrepreneur, leader of the German Society for Mechanics and Optics), Wilhelm Foerster (director of the Royal Observatory in Berlin, professor of astronomy at the University of Berlin, co-founder of the Astrophysical Observatory at Potsdam, director of the Imperial Institute for Weights and Measures, ex-officio member of the Prussian Central Committee of Ordnance Survey, member of the International Bureau of

Weights and Measures[1]), and Carl Bamberg (commercial advisor, entrepreneur, technologist). They were joined by leading mechanics and opticians from throughout Europe. Other dignitaries joined the festivities: city advisor D. J. G. Halske, Prussian minister Heinrich von Boetticher, royal minister of state and finance Adolf von Scholz; minister of the interior Gustav von Gossler; secretaries of state Heinrich von Stephan and Karl Herzog; undersecretary Hermann von Lucanus; director of the ministry Greiff; Bavarian Count Hugo von Lerchenfeld-Köfering; and Bavarian Major General Robert von Xylander. The back wall of the auditorium was bedecked with flowers surrounding a bust of the Bavarian optician. The crowd fell silent as royal music director E. Schultz approached the podium to conduct the Caecilia Choir of Berlin in the singing of the *Fraunhofer-Hymnus*, an anthem whose words had been written in 1831 by Eduard von Schenk and set to music for the occasion by Schultz. The haunting words permeated the hall:

A higher spirit occupied this corpse.
A spirit that is now sitting on the throne of its home.
Because his home was not the Earth,
But the cosmos, and the eternal light:
An eagle, which climbed into the sun
And uncovered the secrets of its light.
The stars followed his powerful call
And the magical glass that he created.
He snatched the stars from the night.
He divided and doubled them
And brought us closer to them.
The distant stars behind the misty veil
Obediently manifested themselves to his telescope
And permitted him to see in this burst of flame
Where the suns formed and died.
He diffracted, divided, measured the ray,
Combined and scattered it according to law and choice.
He held tightly the light of Sirius.
The glimmer of Vega plays in his hand
And as his fame pervaded the world,
Enhanced by flying further and further across the seas,
He remained silent, gentle, humble as a child
Full of naivete and ever devoutly minded.[2]

Teary-eyed Germans with extended chests erupted in thunderous applause. Silence ensued as Germany's leading physicist, Hermann von Helmholtz, approached the podium. After a formal greeting, he explained what was being celebrated:

What we are celebrating is actually a day of remembrance for Germany's middle class (Bürgerthum), to which we proudly have occasion to call everyone's attention. Of all the various orientations of bürgerliche Arbeit, the art of practical mechanics holds an eminent place. . . . Mechanics stands at the top, striving for the highest level of exactness, pureness, and dependability of its work. . . . I myself am one who, through much experience, can provide evidence on how high these first and most important values [exactness, pureness and dependability] of bürgerliche Arbeit have risen with the leading mechanics: how one always finds a master craftsman who has earned the greatest respect—not only in many university cities throughout Germany, but also in some middle-sized cities without a university.[3]

Helmholtz emphasized that Germany respected the labor of the mechanic in general and the Handwerker in particular. He claimed that he had learned much from mechanics, particularly instrument makers. Indeed, he wondered where physics, astronomy, atmospheric research, the navigational telescope, and electric telegraphs would be without "the intelligent assistance of practical mechanics."[4] Helmholtz then turned to the subject of his oration:

This class of Bürger, which in its quiet ways guards and sets in motion the best virtues of German Bürgertum, celebrates today the memory of one of the first and most famous men from its own ranks. He rose from the poorest conditions using his strength and industry, worked against difficult obstacles, and became the owner of the most famous Optical Institute on Earth at the time. His scientific discoveries have expanded our knowledge of the cosmos in ways previously unimagined, and have provided astronomers, physicists, and chemists with the task of completing the studies that these discoveries have sparked.

Fraunhofer has grown up and risen from the soil of the artisan (Boden des Handwerks).[5]

Helmholtz then offered a brief summary of Fraunhofer's life based on Joseph von Utzschneider's account of 1826, underscoring Fraunhofer's working-class origins, his industry, his frugality, and, just as important, the wisdom of his patron, King Maximilian I.

Helmholtz summarized Fraunhofer's Six Lamps Experiment, which had greatly enhanced the accuracy of the measurement of refractive indices, thereby allowing the production of a finer quality of achromatic lenses. The explanation of the nature of the solar dark lines had been a source of much debate and controversy until, according to Helmholtz, the German chemist Bunsen and the German physicist Kirchhoff had determined the relationship between emission and absorption lines, culminating an all-Germanic research program. Helmholtz made Fraunhofer the creator of a long and impressive lineage of German physical scientists and their research. In this respect Helmholtz overreached, because Fraunhofer was in no meaningful sense the forefather of Kirchhoff and Bunsen's research. He had never considered the nature of the dark lines of the spectrum; he was only interested in using them for testing the quality of his lenses.

During the 1870s and the 1880s, German physicists wished to differentiate their physics from its British counterpart. The British physics community figured prominently in Helmholtz's speech. Helmholtz discussed how Fraunhofer had used Newton's discovery of mother-of-pearl colors to create perfectly spherical lenses. Fraunhofer's measurements of the phenomenon had surpassed the Englishman's.[6] Helmholtz even compared Fraunhofer to Michael Faraday, since both were great experimentalists who had made discoveries in physics, and both had come from humble backgrounds. But Helmholtz wished to make a crucial distinction between the two. He claimed that Fraunhofer, unlike Faraday, had remained attached to "the soil of artisanal labor," from which he came.[7] Fraunhofer's scientific achievements had come from his artisanal ability to produce finer achromatic lenses. Faraday, according to Helmholtz, had shed his artisanal, humble background. Hence,

> Fraunhofer is, in this respect, the model that shows us to what heights the labor of the artisan can lead, when a talented man's entire industry, loyalty, and perspicacity are used to remove any deficiencies. It was different with Faraday, when the book bindery merely provided him with an incidental impulse and managed to keep him alive until Humphry Davy accepted him. He never worked on practical assignments; only the desire to research (Forschungslust) filled his breast. Only when his fatherland ordered him to undertake practical research, such as the lighting of the lighthouses and the preparation of optical glass, did he work on technical questions.[8]

Helmholtz then used Brewster's words to illustrate Germany's early-nineteenth-century hegemony in the manufacture of optical instruments. In the *Edinburgh Journal of Science* of 1825, Brewster admitted that the Bavarians had usurped the British optical empire.[9] Helmholtz concluded, however, with the admonition that by the 1850s Britain had rebounded and once again gained world supremacy, and it was up to the kaiser's government to finance research on optical glass by Ernst Abbe and Otto Schott of the Carl

Zeiss Works in order to recapture the lead in optical technology and in the financial markets.

After another round of enthusiastic applause, Helmholtz descended into the audience, and Privy Senior Administrative Official (Geheim Regierungsrat) Prof. Wilhelm Foerster, who strongly supported Schott and Abbe's research on optical glass, ascended to the podium. Once again the audience was treated to a short biography of Fraunhofer featuring the rescue of the young child from the wreckage of a collapsed house as Prince Maximilian Joseph IV looked on. Once again the imagery of light and darkness was employed to contrast Fraunhofer's optical research on light with the darkness of his financial position as an orphaned child. Foerster proceeded to link Fraunhofer's "technology of precision" (Präcisionstechnik) to a more encompassing talent of the German people (Volk). He claimed that when it came to the "rigorous arts" (strenge Kunst)—i.e., precision technology, as opposed to the aesthetic arts—the labor of German men and women stood alongside important German intellectual work.[10] Foerster traced this "talent of the Germanic race" (Begabung germanischen Stammes) back to Fraunhofer. He offered a brief history of optical instruments that depicted England as the undisputed leader up until the beginning of the nineteenth century, or until Fraunhofer appeared on the international scene. Foerster underscored not only that Fraunhofer's telescopes were being employed in observatories all around the world, but also that Fraunhofer's work had given a great boost to Germany's mechanical and optical research. Fraunhofer became the emblem of German precision measurement, and mechanical and artisanal knowledge.

Echoing Helmholtz's concerns, Foerster concluded by warning his audience, particularly the numerous government officials present, that the English and the Americans were then improving optical technology beyond Fraunhofer's work. Germany had begun to slip back into its eighteenth-century obscurity.[11] He called attention to Schott and Abbe's work on perfecting achromatic lenses, and urged the Prussian government to resume its funding of their work.

After Foerster's speech, the choir again broke out in song. An ode had been composed for this event in honor of German mechanics:

Rugged was your youth, as the mountain current
Wildly bursting through the rock bed: only effort and work.

But your genius cleared for you
The path to eternal fame.
You based the theory of light
Firmly on the honorable foundation of practice (Praxis).
To the most distant lands you carried the reputation,
German Mechanic.[12]

At the song's completion, the president of the German Society for Mechanics and Optics, Fuess, announced the establishment of a foundation in memory of Fraunhofer that would support young German mechanics and opticians in theoretical and practical training. Fraunhofer served as the model for many young Germans and foreigners alike. The governing body of this foundation would be composed of leading mechanics from all parts of the German Empire so that "the general character of the nation will clearly emerge."[13] Fuess proclaimed to his colleagues that "Fraunhofer was one of us" and congratulated the Berlin city government for its wisdom in establishing a technical college (Fachschule) exclusively committed to the education of young mechanics and opticians.[14] After Fuess's announcement of an endowment of 11,000 Reich marks with a 400 Reich marks annual increase, and a plea for the society's assistance in keeping the foundation thriving, the choir concluded the ceremony with a familiar anthem that captured the essence of the German spirit: Beethoven's "Die Himmel rühmen des Ewigen Ehre!"

Hermann von Helmholtz

Helmholtz was undoubtedly the most influential German scientist of the 1870s and the 1880s. He was one of Germany's leading "bearers of culture" (Kulturträgern) and representatives of the "educated upper-middle class" (Bildungsbürgertum), and he had assumed the role of spokesman for Germany's scientific elite.[15] His speech at Fraunhofer's anniversary celebration must be viewed in the context of his support for the Physikalisch-Technische Reichsanstalt, whose approval was to be voted upon by the Reichstag less than three weeks later.[16] The status and importance of skilled artisans such as instrument makers were being challenged during the Kaiserreich, as the influence of the working class was on the decline. Skilled artisans were being "industrialized" and "proletarianized" much as they had been in Britain 50 years earlier. This perceived decline spilled over into the debate as to whether instrument makers should be admitted to the PTR.

Figure 8.1
Hermann von Helmholtz (1821–1894). *(Deutsches Museum, Munich)*

Helmholtz emphatically supported their inclusion, and the hundredth anniversary of Fraunhofer's birth served his purpose perfectly. He repeatedly underscored Fraunhofer's humble origins and his ability to overcome a life of poverty and join the Bürgertum. Fraunhofer, Helmholtz argued, came to epitomize all that was great about the German middle class (Bürgertum). But where Helmholtz's oration is most interesting is its emphasis on the importance of skilled artisans to Germany. Such importance, Helmholtz argued, could be historically demonstrated. The PTR was to become, he hoped, a late-nineteenth-century example of Fraunhofer's Optical Institute. The technical section's artisans (instrument makers and precision mechanics) were to work hand in hand with scientists from the science section. This cooperation would result in a successful scientific, technological, and entrepreneurial venture, as had been the case with the Optical Institute. This was his message to the numerous members of the Reichstag present at the Fraunhofer celebration.

As Cahan has suggested,[17] the genre of public lectures enabled Helmholtz to emphasize a pan-Germanic culture—a culture with which all Germans could identify. Germany, unlike either France or Britain, was until 1871 composed of independent states, each with its own culture. Power was very decentralized. In terms of cultural significance, Berlin never was another Paris or London. Also, Germany did not have a national religion; it was divided, state by state, between Roman Catholicism and Lutheranism. Bitter rivalries between Protestant Prussia and Catholic Bavaria became commonplace during the second half of the nineteenth century. Interestingly, Helmholtz sought to base cultural unity on scientific prowess rather than on the German language. Such prowess resulted, in part, from Germany's unique ability to incorporate artisanal labor into the scientific enterprise. Not coincidentally, he claimed that neither France nor Britain could achieve this. Echoing Rudolf Virchow's notions of the role of science in the Reich (discussed below), Helmholtz believed that science and technology would bind Germany's diverse cultures to form a stronger nation.

Finally, Helmholtz's orational history of Fraunhofer had yet another purpose. Helmholtz saw his history as the final step in a fifty-year process whereby German hegemony in the sciences could finally be announced to the world. Within 50 years, Germany had transformed its science from an embarrassment into world preeminence. Hagiographical history writing is the culmination of a discipline's domination. Its purpose is clearly political. Helmholtz's history proclaimed that German scientists now enjoyed a first-rate status. Helmholtz's public lectures, including his account of Fraunhofer, though targeted primarily at Germany's rulers and its educated middle class (Bildungsbürgertum), were also intended to enlighten Europe's political and social elite.

Ernst Abbe

The Berlin celebration was one of several held throughout Germany to commemorate the hundredth anniversary of Fraunhofer's birth. The Munich chapter of the German Society for Mechanics and Optics co-sponsored the Munich festival with city officials, while the Physics Association (Physikalischer Verein) in Frankfurt-am-Main offered its own day of remembrance.[18] And on 5 March 1887, a day before the Berlin pageant, Abbe delivered a speech in the auditorium of the Physics Institute of the University of Jena.[19] Although the pageantry of Jena never could quite rival that of Berlin, Abbe's oration was enlightening. He, like Helmholtz, claimed that "under his [Fraunhofer's] impulse, a highly developed technical art arose, which still serves today as an unparalleled model of the inner collaborations of a pure science and practical skill (praktischer Geschicklichkeit)."[20] Fraunhofer's relentless endeavor and "skilled hand" (geschickte Hand) resulted in the perfection of a method that made "research into the secrets of nature" possible.[21]

Like Helmholtz, Abbe offered his listeners a brief history of Fraunhofer's life. "As so many of our important men, Fraunhofer came from the soil of

Figure 8.2
Ernst Abbe (1840–1905). *(Deutsches Museum, Munich)*

our people, from which the archetypal and unweakening strength (Kraft) continually renews the driving forces (Triebe) responsible for the intellectual blossoming of the nation."[22] Once again, the importance of the skilled artisan is stressed. But in Abbe's account, Utzschneider played a much more prominent role than in Helmholtz's account. The audience was informed that Utzschneider was a man of extraordinary and multifarious talent, education, and sophistication who possessed an entrepreneurial spirit (Unternehmungsgeist) and a first-class organizational talent.[23] Utzschneider's entrepreneurial ability and his skilled and daring organizational maneuvers both recognized and coordinated the labors of Fraunhofer, the instrument maker Georg von Reichenbach, and the skilled artisan and watchmaker Joseph Liebherr. That coordination was precisely what was needed to restore Germany's cultural commitment to and achievement in the mechanical arts (mechanische Kunst).[24] The importance to the Bavarian enterprise of Pierre Louis Guinand, the Swiss bell- and clockmaker who had taught Fraunhofer the art of manufacturing optical glass, was generally ignored. Although Abbe (unlike Helmholtz) mentioned Guinand's work with achromatic lenses, he did so in a derogatory manner, complaining that Guinand had not improve the manufacture of optical glass beyond the level of mediocrity and casting doubt on his perceived success.[25] Abbe specifically asserted that Fraunhofer, not Guinand, had been the first to succeed in manufacturing glass for achromatic lenses.[26]

The Bavarian trio Utzschneider, Reichenbach, and Fraunhofer had reformed the mechanical arts. This reform had led to astronomical discoveries and to advances in various disciplines that required precision measurement. The audience was told how the labor of these three men, but Fraunhofer in particular, had enabled the exciting research of German astronomers, such as Friedrich Georg Wilhelm Struve and Friedrich Wilhelm Bessel.[27] Thanks to the work of these three, the mechanical arts had returned to Germany after centuries of absence. No longer did Germans need to turn embarrassingly to the instrument makers of the capital cities of their rival nations, France and Britain.[28] Since the beginning of the second half of the century, German industry could extol their unchallenged predominance in instrument making.[29] Munich, according to Abbe, had become the capital of instrument builders for the entire scientific world. In short, a recovery felt throughout Germany in the mechanical arts was the result of the success of Fraunhofer's Optical Institute. And since that

Figure 8.3
Fraunhofer (with his prism spectral apparatus), Georg von Reichenbach (bent over peering through the apparatus), and Joseph von Utzschneider (seated left). *(Deutsches Museum, Munich)*

day, Germans had enjoyed a world-recognized lead in all branches of scientific industry. Fraunhofer's work was the advent of the "exact arts"[30]:

> If justifiable pride in the success of our fellow citizens, as a result of their successful participation in the promotion of broader cultural interests, is an undoubtedly good and worthwhile form of national pride and national ambition, then one may find gratification in such unrestrained expression at a commemoration of Fraunhofer. Through him new paths have been found and paved for our people, paths upon which we can bring to bear with honor our natural gifts and the advantages of our raised standard of education to the peaceful competition of nations.[31]

Abbe bemoaned the fact that the artisanal knowledge of achromatic-lens production had been buried with Fraunhofer in 1826. That knowledge, according to Abbe, could never be recovered. He even asserted that 50 years later much that was being done by Fraunhofer's followers in the production of optical glass had actually been invented by Fraunhofer.[32] It had taken the labor of the two generations following Fraunhofer's death to offer new solutions to the problems of achromaticity.[33]

Abbe and Helmholtz created the myth of Fraunhofer as the forefather of spectroscopy—a field in which German domination had been apparent ever since Kirchhoff and Bunsen had provided the theoretical explanation of the absorption lines and their relation to emission lines.[34] The crowd was treated to a detailed account of how Fraunhofer's research interests predominantly dealt with practical problems that were solved by a combination of a high degree of skill and scientific theory.[35] Abbe concluded his oration by discussing his and Otto Schott's research in the production of apochromatic lenses in the 1880s at the Carl Zeiss Works. By appealing to Germany's history of recognizing the importance of the artisanal knowledge of the mechanical arts, he was hoping to encourage governmental officials present at his oration to continue to subsidize precision mechanics in Germany. A brief history of Abbe's work is necessary in order to appreciate his contribution to the Jena celebration of Fraunhofer.

Abbe was Germany's foremost expert in optical instrumentation during the late nineteenth century. He was born to a working-class family in Eisenach, Thuringia, on 23 January 1840. His father had financed his education with frugal economizing, money earned from Ernst's tutorials in math and physics, and the cash Ernst had won from science competitions at his Gymnasium. Ernst studied at the Universities of Jena and Göttingen, finishing his doctoral exam at Göttingen at the age of 21 with such an impressive result that he was asked by Wilhelm Weber and Bernhard Riemann to become the assistant in Göttingen's renowned observatory. After teaching at the Physics Association in Frankfurt for a year and a half, the 23-year-old Abbe returned to Jena.[36]

Politically, Abbe remained liberal throughout his lifetime. He often recalled how, at age 8, he had witnessed his father secretly house socialists who were fleeing from the guns of the Prussian militia as it attempted to crush the unsuccessful revolution of 1848. In 1859 Abbe joined the German National Association (Deutsche Nationsverein), a liberal political party striving for democratic, peaceful unification of Germany.[37] Such a view was very much in line with that of his employer, Carl Zeiss. Zeiss, owner of what was to become the world's leading supplier of optical equipment, needed a physicist well acquainted with optical theory and mathematics. Abbe was the perfect choice, and he joined the company in 1866. Although it would be beyond the scope of this book to discuss Abbe's research on apochromatic lenses,[38] it is necessary to uncover why Abbe used Fraunhofer as a historical resource.

Zeiss and Abbe's cooperation was predicated upon the unification of science, technology, and entrepreneurial spirit. Abbe's goal was to create a business where the Pröbeln (trial-and-error procedures) of previous opticians could be replaced by scientific principles, thereby ensuring replication of artifacts. Abbe complained that these artisans did not have a rule-governed method of manufacturing optical glass. According to Abbe, such a method could ensure optical quality. By the late nineteenth century, Abbe had amassed a large amount of capital for both the Carl Zeiss Company and himself. He became one of Germany's foremost entrepreneurs during the late nineteenth century and the early years of the twentieth. But despite his capitalist corporate ideology, reform of workers' rights and working conditions preoccupied him until his death in 1905. Much of Abbe's entrepreneurial point of view concerned the well-being of the skilled artisans at the Carl Zeiss Works.

Abbe's historical account of Fraunhofer underscored Fraunhofer's humble beginnings and cooperation with Utzschneider in establishing the Optical Institute, which became the envy of the scientific world. The entrepreneurial tradition, Abbe was arguing, was crucial to the formation of a late-nineteenth-century optical precision technology—just as crucial as the craft tradition epitomized by Fraunhofer. Interestingly, Abbe was embroiled in debates about the role of the technical section of the PTR during precisely this period. He argued, *pace* Werner von Siemens, that instrument makers should not be used by industrialists simply to undertake routine tests for standards and quality control, but rather should be permitted to have a greater effect on the direction of research at the PTR. Hence, Abbe underscored the relationship among Utzschneider, Reichenbach, and Fraunhofer: the managerial and craft traditions were both necessary for precision optical research.

Abbe clearly had vested interests in his construction of Fraunhofer. First and foremost, his history sent a clear message to Prussian officials present at his oration that governmental subsidy of science and technology had a historical precedent on German soil. Abbe praised the foresight of King Maximilian I's patronage of the skilled Bavarian artisan. Germany could recapture world hegemony in optical technology if, and only if, Prussian officials would resume their patronage of the Carl Zeiss Works, which had been ended in 1886. Second, Abbe saw himself unifying technical knowledge with scientific theory, which was precisely what he claimed Fraunhofer had been doing earlier that century. Third, the importance of German skilled artisans to the German scientific enterprise had been acknowledged some 50 years earlier. As Germany was chasing Britain and France for world supremacy in scientific and technological industries, once again it was obvious that opticians, precision mechanics, and other skilled workers associated with those industries were prized possessions for the new Reich. That recognition, which was under threat during the Kaiserreich, would propel German industry past their rivals just as the labor of skilled artisans, under the direction of entrepreneurs, had assisted German science in reaching the forefront of scientific accomplishment.

The Entrepreneurial Moral of Abbe's Account of Fraunhofer

A large part of Abbe and Schott's enterprise was aimed at playing upon the aforementioned chauvinistic sympathies in the newly formed Reich. They argued to Prussian government officials that Germans had previously been the status of lensmakers to the world. Fraunhofer, with the financial assistance and sagacity of King Maximilian I of Bavaria, had produced the world's finest achromatic lenses during the second and third decades of the nineteenth century. Britain and France, as well as the rest of the "civilized world," turned to the German artisan (Abbe and Helmholtz emphasized that Fraunhofer was German rather than Bavarian) and his Optical Institute for their astronomical and surveying instruments. Abbe and Schott hoped that the Prussian government would continue to emulate Bavaria's earlier patronage of the sciences.

On 30 March 1882, Abbe and Schott furnished the government officials in the Prussian Imperial Ministry with a report on their experimentation on the improvement of optical glass.[39] Both men asserted that the art of practical optics had not experienced any theoretical advances since the work of Fraunhofer. Improvements in glass production had been restricted to chemical manipulation provided by the "usual bürgerliche glass technology."[40] They bemoaned the fact that even modern British and French glass manufacturing was based on purely commercial interests and on empirical, trial-and-error methods:

> Fraunhofer seems never to have existed for the present representatives of this industry. How can such an industry, whose entire operation has been founded upon a minimal amount of scientific work, be capable of pursuing their assignments, since they all are based upon the framework of the old [pre-Fraunhoferian] tradition?[41]

Abbe and Schott argued that they had built upon and advanced Fraunhofer's combination of optical theory and chemical manipulation of glass manufacturing. By employing the skilled labor liberally provided by the Carl Zeiss Works in Jena, Schott had begun his glass experiments in the beginning of 1881.[42] In November of 1881 both men decided to erect a laboratory in Jena in order to investigate the properties of optical glass.[43] By March of 1882, they had carried out more than 100 melting probes, measuring the refractive indices of the spectrum between Fraunhofer lines *A* and *G* with unprecedented accuracy.[44]

Although both men were convinced that the financial and technological support provided by Carl Zeiss would suffice for the completion of their present experiments, they requested Prussian subsidies for further studies and for scaling up their enterprise. It is important to note that Jena is located in the German state of Thuringia, not in Prussia, and that Thuringia was comparatively poor. Hence, Prussia's subsidy would be based on the importance of Abbe and Schott's optical glass to optical companies in Prussia as well as on Prussia's interest in setting the example for scientific and technological research in the Reich. Schott and Abbe envisaged their type of research as playing a pivotal role within Foerster's planned institute for precision mechanics.[45] They supported their application by claiming that their research was crucial for the scientific industries throughout the entire Reich—particularly those involved in manufacturing glass, the largest of which was the Rathenow Optical Company in Prussia.[46] Abbe and Schott argued that state support would permit German industry to free itself from its "dangerous dependency" upon foreign materials and would stimulate the German economy.[47] German science, technology, precision mechanics, and industry would all benefit. They concluded by playing on Germany's greatest fear: that without intervention by the Prussian government, their preliminary research would benefit the British and the French, and German industry would be left with only "the tidbits."[48] In order to strengthen their request for governmental support, they enrolled the support of Foerster, who composed a memorandum recommending their research without reservation (and who, recall, had been at the Fraunhofer celebration in Berlin in 1887).[49] This letter was submitted with Abbe and Schott's original report. During the early 1880s, Foerster sought to get the state to subsidize an institute for precision measurement that would employ skilled instrument makers—precisely the kind of institution where Abbe and

Schott wished to conduct their optical research.[50] Indeed, Abbe and Schott's report of 30 March 1882 must itself be understood as an offer of support. Both men were very sympathetic to Foerster's attempt to procure funds from the Prussian Ministry with a view to foster the work of skilled artisans affiliated with science and technology, particularly optics.[51]

During the second half of the nineteenth century, several Prussian intellectuals feared that the German territories' instrument-making capacity was in reality quite negligible. By the second half of the nineteenth century, craftsmanship, which had always trumped the fledgling German industrial system, was finally succumbing to the organized factory system.[52] Skilled artisans were losing their privileges, much as they had in Britain nearly 50 years earlier. Mechanics and makers of scientific instruments were included in this decline. With the backing of the Berlin Academy and the Prussian Ministry of Education, Karl Schellbach assembled a commission to study the competence of Prussian precision technology. The Schellbach Commission completed its study in 1872, concluding that Prussia's instrument makers lagged woefully behind France's and Britain's. In 1873 the Prussian Academy of Sciences publicly responded to the Schellbach Commission's proposal: it refused the commission's request to divert funds for the training of instrument makers. The academy argued that the ministry should direct funds for the development of precision technology toward existing institutions rather than the Academy.[53] The Academy frowned upon the commercial orientation and practical application of the work of artisans affiliated with precision mechanics.[54] Indeed with regard to the Academy's stance on skilled craftsmen, not much had changed since Fraunhofer. Foerster, who in the 1870s became Berlin's director of education, disagreed. He quickly established a military commission, under the directorship of Prussian army general Otto von Morozowicz. This commission, like Schellbach's, concluded that Germany's instrument makers trailed dangerously behind France's and Britain's. Foerster and his allies then moved to establish a Berlin institute for precision mechanics and optics, modeled after the Conservatory of Arts and Crafts (Conservatoire des arts et métiers) in Paris, where instrument research could be carried out and future instrument makers could be trained. Such an institute was precisely what Abbe and Schott wanted to see established.

As Terry Shinn has argued, in the 1870s and the 1880s Berlin's instrument craftsmen prodded the new imperial government into recognizing the nation's need for a strong precision instrument capacity, since instrument makers were clearly on the decline during the early Kaiserreich.[55] Throughout 1874 and 1875, Foerster ran into staunch opposition from the Berlin Academy and even from circles within the Ministry of Education, who argued that the skills possessed by instrument makers fell into the domain not of science but rather of engineering and industry, and hence they should not be the financial source for Foerster's institute. Foerster countered by claiming that instrument makers were crucial to both. Leopold Loewenherz sided with Foerster in the debate. As a leading member of Germany's Imperial Institute for Weights and Measures, Loewenherz fought from the late 1870s onward to establish a specialized institute. Indeed, he became the co-founder of the German Society for Mechanics and Optics with the goal of surpassing the achievements of British and French instrument makers, and he was present at the Fraunhofer celebration in Berlin. Hence, it is in this context that one begins to understand Foerster and Loewenherz's support for Abbe and Schott's apochromatic-lens enterprise. Foerster echoed Abbe and Schott's argument for state subsidies. He cautioned the government that the British scientists W. V. V. Harcourt and G. G. Stokes had been attempting to solve similar problems.[56] (Harcourt had begun experimenting with glassmaking, which was to occupy much of his time until his death in 1871, in 1834. Stokes subsequently joined him, and both pursued the inquiry initiated by Fraunhofer and Faraday concerning the effect of chemical composition on the distribution of dispersion with the goal of producing achromatic glass. By the 1860s, Harcourt and Stokes had made several significant breakthroughs concerning the influence of particular substances (such as boric acid) on dispersion.) Having argued that the Prussian government should provide financial backing so as to secure German science and industry, Foerster concluded his letter by restating the need for an institute for precision mechanics.

Carl Bamberg, owner of one of Germany's foremost workshops for precision scientific instruments, added his support for Abbe and Schott.[57] In a letter dated 7 October 1883, Bamberg proclaimed that Schott had repeatedly produced optical glass 12–15 centimeters in diameter of a quality that neither Chance nor Feil could achieve.[58] Schott had surpassed Fraunhofer's high standards: among other innovations, he made it possible to replace the objectives of achromatic telescopes without rebuilding the telescopes. Bamberg's letter is informative because it too discussed the concerns of

German entrepreneurs and governmental officials. He reported that Schott had informed him personally that if the Prussian government did not intervene quickly and subsidize his research he would be forced to turn to either the Chance Brothers or Feil for support, and he would be warmly welcomed.[59] Bamberg continued his admonition:

> Just how depressing the loss for Germany would be in terms of the nation's scientific, technical and commercial conditions, I would prefer not to mention. It would represent a sad page in the history of the German art of precision, which would record how such an undoubtedly important achievement was relinquished to a foreign country. England, quite famously, had to pay the penalty for a similar occurrence for four decades. The work of Dollond and Ramsden was quickly eclipsed. The English paid much money to Fraunhofer's Institute in order to purchase good telescopes.[60]

Bamberg, who was also present at the Fraunhofer celebration, was arguing that there had been a historical precedent for the failure of a government to support optical research. As we have seen, the British government had hindered optical research in the late eighteenth century and the early years of the nineteenth. Their optical technology had been surpassed by Fraunhofer, and as a result the British had lost a significant source of revenue as well as their optical monopoly. Bamberg used history to legitimize governmental support of the scientific enterprise. The lobbying on behalf of Abbe and Schott was successful; the Imperial Ministry agreed to pay the two men 25,000 marks for the fiscal year 1884–85.[61]

Less than two years after their successful plea, Abbe and Schott reviewed their results.[62] Although their experiments were coming along nicely, more support was necessary. Both men decided that an experimental station for glass should be built in Jena, in the vicinity of the Zeiss Works.[63] This work station was not meant to replace the existing laboratory; rather, it was intended to become a space for researching the technical aspects of glass manufacture and for instructing working assistants. Although it was hoped that the Prussian government would assist in the laboratory's construction, all laboratory organization, training, and profits were to be solely in the domain of the Carl Zeiss Works. The men provided a detailed design of the experimental station, as well as an estimated cost of 37,950 marks for its erection and maintenance for the first year.[64] Once again, foreign competition was mentioned. For several decades, the manufacture of optical glass had been monopolized by two major scientific entrepreneurs Chance (in Birmingham) and Feil (in Paris). Over 50 years, these men had established and perfected a routine for manufacturing glass. Abbe and Schott had much ground to cover.[65] Feil and Chance, however, were motivated entirely by commercial and materialistic interests; therefore, Abbe and Schott argued, the art and spirit of precision mechanics had been lost ever since Fraunhofer's death.[66] Abbe and Schott intended to bridge the huge chasm between the optical glass industry and the scientific and technological advances of the previous 50 years. Their research could, in the end, lead to the independence of German industry from foreign companies.[67] This independence was particularly crucial for the Rathenow Optical Company. This factory, which employed more than 4000 Germans, manufactured field telescopes for the German army and navy. Unfortunately, its raw materials—particularly the optical glass—were imported from Paris.[68] Enclosed in their report was a reference from the representative of the Rathenow industry, commerce advisor Emil Busch, who claimed that a deficiency in Germany's sources of optical glass impeded the development of optical technology in the Reich.[69]

Busch's reference on behalf of Abbe and Schott, dated 27 November 1883, offered a brief yet enlightening history of the manufacture of optical glass in Rathenow.[70] The Rathenow Optical Company had begun in 1800 under the directorship of a Lutheran pastor named Duncker. Its first 45 years had been devoted exclusively to manufacturing reading glasses (and their frames) and magnifying lenses. In 1845 the company had begun to diversify, manufacturing field telescopes, telescope tubes for terrestrial and astronomical purposes, photographic objectives, opera glasses, microscopes, prisms, and military telescopes. The crown and flint glass for the objectives of the telescopes and microscopes and the fine optical glass for the prisms had to be imported from France and Britain. Busch then proceeded to offer a history (a rather accurate one, as it turns out) of the Optical Institute in Munich, which emphasized Fraunhofer and the Swiss Pierre Louis Guinand. He recited the account of how Guinand's son, Henri traveled to France and revealed his father's secrets to Georges Bontemps, who later passed on the information to Chance Brothers of Birmingham while employed there. He also told of Henri's collaboration with his grandson, Charles Feil, who collaborated with his son Edmond to establish one of the finest optical companies in Europe. France's and Britain's successes, Busch explained, had their roots in German soil, and, ironically, Fraunhofer was the source of their success.

Although Feil was one of the world's best opticians, Busch claimed that he lacked a "scientific" method of producing optical glass. Busch echoed Abbe and Schott's assertion that British and French optical glass manufacture was based on trial-and-error methods. Apparently, Feil had once produced a batch of high-quality flint glass for photographic purposes, but could not replicate his success.[71] Therefore, Busch argued, optical glass had to be manufactured by individuals who possessed a thorough knowledge of chemistry and theoretical optics.[72] Although theoretical knowledge was necessary, it certainly was not, Busch was convinced, sufficient. In 1869, Busch paid one of P. L. Guinand's great-grandsons to come to Berlin and assist him. Guinand's descendent, we are told, had witnessed the actual practices of glassmaking and understood the theory, but had never actually made glass himself.[73] He failed to produce glass of the same quality as his great-grandfather. Baron von Bredow (from Wagenitz, near Rathenow) had the same experience. Both men lost more than 20,000 taler in capital.[74]

Busch pointed out to the Prussian government that this was an opportunity for German industry to improve its international credibility, as German science had already done. Indeed, Germany's scientific and technological industries could exploit the reputation of Germany's scientists. But in order for that to happen, the Prussian government had to intervene. Busch complained that strict German labor laws did not allow fair competition, particularly when compared with the French. For example, Morez, a French manufacturer of reading glasses, produced high-quality glass at very cheap prices by employing women and children at scandalously low wages.[75] Although he was not arguing against Prussia's labor laws (he praised the fact that women and children were not permitted in the workhouses), he did feel that government subsidies should be given in order to reduce the price of reading glasses manufactured in Rathenow, thereby leveling the playing field. The government also needed to intervene since French commercial practices, according to Busch, were unfair. The Rathenow Optical Company imported crown and flint glass from France in order to produce field telescopes and opera glasses. Busch complained that the French optical glassmakers gave preferential treatment to the only other Continental competition, the French company Lemaire, by supplying them with better glass at a much lower cost. He spoke of France's nationalism and mistreatment of German companies. "Nowhere in the world," he concluded, "does such a partiality rule as in

France, especially in Paris."[76] Busch also asserted that the French were prejudiced against any German-manufactured product.

After demonstrating how crucial it was for the Prussian government to assist in the creation of a factory for raw optical glass within Germany for scientific and economic reasons, Busch concluded with perhaps the most compelling reason of all: war. Should war break out once again between Germany and France, shipment of optical glass for army and navy field telescopes from France would obviously be halted. Also, there was always a chance that the Feil Company would be shelled, as it nearly was in 1870. In either case, the German optical industry would be dealt a fatal blow. The startling realization of Germany's dangerous dependency upon foreign glass had also occurred to the German navy. During the late nineteenth century when Germany began building its naval fleet, Hamburg, Bremerhaven, and Rostock were vital port cities for both trade and defense. Yet Germany had to import lighthouses, preassembled, from France and Britain.[77] German Fresnel (zone) lenses were of inferior quality, and the British and the French would not ship raw glass; they would sell only the final product.[78]

Along with their January 1884 report, Abbe and Schott sent a quality comparison of their lenses with lenses obtained from the Chance Brothers in Birmingham,[79] based on spectrometric measurements made by Abbe and his assistant P. Riedel.[80] Their strategy was rather straightforward. Since they purported that the optical glass supplied by Chance Brothers was the model by which all optical glass manufacturing was to be judged, if their own products could remedy chromatic aberration more effectively (i.e., by correcting more rays), then it would, de facto, set the new standard for optical precision. They recreated their opponents' achromatic lenses by combining two lenses out of the seven that Chance Brothers had sent them. Four combinations of British flint and crown glass yielded the best lenses. Those were the samples whose standards were to be surpassed. Abbe and Riedel noticed that the dispersion of the crown-and-flint-glass combination increased significantly when moving from the red to the blue end of the spectrum. Such a difference in dispersion results in the so-called secondary spectrum (chromatic aberration) that inevitably arises from combinations of crown and flint glass.

The Jena glass samples, which were produced before Abbe and Riedel's analysis of the foreign glass, however, did not exhibit a general increase in dispersion from red to blue. Indeed, the undesired dispersion was at a minimum in the interval from *Na* to *Hγ* (Fraunhofer lines *D* to *F*), the portion of the spectrum where the human eye is most sensitive. These lenses corrected three colored rays of light: the extreme red and blue, and the yellow rays from the middle of the spectrum. Their lenses were superior to their competitors'. In sharp contrast to British attempts to replicate Fraunhofer's lenses during the third decade of the nineteenth century, Abbe and Schott were able to produce lenses of a higher standard than Feil's and the Chance Brother's without witnessing their competitors' labor practices. Schott's chemical knowledge of optical glass, combined with his and his assistants' knowledge of optical-lens production, were sufficient for them to proclaim that they "had completely mastered the technical conditions of the production [of Feil's crown extra-dur and Chance's flint double-extra-dense] and [knew] how each glass of these types must be handled in order to achieve a high level of lens production based on the current standards."[81] Schott and Abbe were indeed able to reverse engineer and improve upon the lenses of Feil and the Chance Brothers.

Schott and Abbe's report of January 1884, which was sent to the Ministry of Culture, was just as successful as their original report of 1882 had been. The Optical Industrial Institute of Rathenow added its support by writing a testimonial to Loewenherz encouraging the development of an optical glass plant on German soil. The members of that institute appreciated the importance of high-quality glass in solving the new needs of the Reich; indeed, it is clear that they were aware of the demands of the kaiser's admiralty for Fresnel lenses to be used in lighthouses. These lenses had to be totally homogeneous, free of bubbles and striae, and not easily oxidized.[82] The raw glass for these lenses could not be produced by German glass factories and therefore, as stated above, had to be imported from Britain and France.[83]

Gossler, head of the Prussian Ministry of Culture, wrote that he was pleased to report that the state budget 1884–85 had allocated 25,000 marks for defraying the cost of constructing a glass factory in Jena. This factory, Gossler continued, would "render German industry in this area [of commerce] independent of foreign competition."[84] On 18 June 1884, Gossler sent a letter to Royal Minister of State and Finance Scholz informing him of the agreement to pay Abbe and Schott 25,000 marks. He informed Scholz that such a subsidy was well deserved since the Jena scientists' work had received unreserved appraisals from Helmholtz and Foerster, whose reputations were unquestionably sound.[85] Gossler was kept informed of the development of Abbe and Schott's progress by frequently submitted reports.[86] Gossler and the Prussian government must have been pleased. On 19 September 1884, the government agreed to increase their subsidy for the fiscal year 1885–86 by 10,000 marks to a total of 35,000 marks.[87] Gossler was one of the most powerful members of the kaiser's government who supported the interactions between science, technology, and industry for the benefit of the Reich. In January of 1887, for example, he informed the Reichstag that the PTR was to be emblematic of a "German national character." He continued: "We are forming in the German Reich . . . a unitary circle of interests in the field of science, trade, industry and in many other fields as well."[88] Hence, not surprisingly, two months later he too attended the Fraunhofer celebration.

In short, Abbe and Schott were able to convince the Prussian government that it should subsidize an optical industry in Thuringia for the improvement of the scientific, technological, and economic status of the Reich. It should be noted, however, that the Prussian government's subsidization of the Carl Zeiss Works lasted only 2 years, ending in 1886. Hence, Abbe's commemoration speech on Fraunhofer of 1887 gave him a perfect opportunity to address the importance of governmental intervention. The importance of science to the young Reich was undoubtedly crucial for its development. Science could stimulate its economy, which could then begin to rival Britain's and France's. But equally important to the Reich were the skilled artisans associated with these scientific and technological ventures. From shortly after Fraunhofer's death in 1826 until just before Schott and Abbe's collaboration in the 1880s, the German territories had been forced to import glass from the Chance Brothers of Birmingham and from Feil of Paris. But by the late 1880s Germany's Carl Zeiss Works was the leading supplier of precision optical instruments and lenses, as a result of Abbe's knowledge of theoretical optics, Schott's knowledge of glass chemistry, and the financial resources of the Zeiss Works and (to a lesser extent) the Prussian government. Schott's Jena glassworks produced more than eighty types of glass, some of which were used in the construction of apochromatic lenses designed by Abbe.[89]

By 1900 the optical industry had become emblematic of the Reich. There were 347 optical firms in Germany, employing some 10,936 employees.[90]

Of those firms, 92 were optical workshops (as opposed to merely mechanical workshops) where optical glass was produced, polished, and cut for astronomical and ordnance surveying instruments and for reading and opera glasses.[91] In 1900 Germany was the world's leading exporter of cut and polished optical glass and instruments, with Great Britain, France, the United States, and Russia as its chief customers.[92] Of the entire production of German optical instruments and supplies in that year, between two-thirds and three-fourths (or approximately 20 million marks' worth) flooded the market.[93] Germany exported nearly 12 million marks' worth of astronomical and optical instruments, over 3.5 million marks' worth of cut and polished optical lenses (including reading and stereoscopic glasses), and over 300,000 marks' worth of raw optical glass (crown and flint) in 1900.[94] The Carl Zeiss Works alone sent 633,000 marks' worth of optical equipment to Great Britain.[95] The company became a symbol of Germany's scientific, entrepreneurial, and technological might.

Scientists Unifying the Reich

During the late nineteenth century, German scientists in general, and German physicists in particular, enjoyed an enviable position among the world's scientific elite. Helmholtz, Abbe, Zeiss, Heinrich Hertz, Kirchhoff, Bunsen, Emil DuBois-Reymond, Ernst Haeckel, Rudolf Virchow, Wilhelm Ostwald, and Werner Siemens all contributed to the scientific success of their infant nation. After centuries of looking westward to France and Great Britain, Germany was at last a nation with a community of scientists who set the pace. The unification of the German territories to form Germany in 1871 and the formation of a united, world-leading scientific community were certainly not coincidental. Indeed, scientists played a part in paving the way for both cultural and political unity. During the late 1860s and the 1870s, unification was being spoken of at epistemological, political, and cultural levels. German scientists actively participated in the unification of scientific disciplines and standards and in their Reich. Helmholtz, for example, unified physiology and physics by reducing life processes to physical ones. There was no room for vitalistic forces in Helmholtz's physiological schemes.[96] Unification was not, however, limited to the reduction of one science to another; it also entailed the combination of previously distinct disciplines. Abbe's work on the physics of apochromatic-lens production, for example, united precision mechanics with optical theory. Both

Helmholtz and Abbe underscored the importance of unifying physical theory with the practice of precision mechanics and instrument makers. Kirchhoff and Bunsen's work on absorption and emission lines united analytical chemistry with physics.

There was also unification of scientific and technological standards within both Germany and Europe. During the 1850s and the 1860s, Germans had realized that, in order to compete in worldwide business markets and to be recognized in the predominantly French and British international scientific and technological circles, they had to unite their own efforts. In 1866 the German states, excluding Prussia, met to negotiate national standards for weights and measures.[97] As the scientific enterprise became more cooperative and internationally oriented, and as trade between European nations increased, scientific and technological standards were needed. Hence, in 1875 the European nations formed the International Bureau of Weights and Measures. This decade witnessed the standardization of electrical units, as evidenced by the first International Electrical Congress in Paris (to which 22 nations sent 250 delegates).[98] It was here that the German contingent—Foerster, Helmholtz, Friedrich Kohlrausch, and Gustav Wiedemann—expressed concern over French and British codominance of metrology. The Germans realized that if they were to prevent foreign supremacy they would need institutional support for metrology within the new Reich.[99]

During the latter half of the nineteenth century, Germans became ferociously nationalistic in cultural affairs, and science was no exception. Indeed, since German science assisted in unifying Germany and its culture, German scientists often held chauvinistic views of the scientific enterprise. The francophobic speech that DuBois-Reymond delivered in his capacity as dean of the University of Berlin in August of 1870, in the midst of the Franco-Prussian War, was the rule rather than the exception. Six months later he reminded the Prussian Academy of Sciences to "uphold the flag of science" while their students, sons, and brothers carried the German banner on the battlefield against France.[100] The pathologist and politician Rudolf Virchow envisaged the scientific enterprise as providing the fledgling nation with spiritual and cultural unity. That is precisely what he meant by the "national meaning" of science. Science could contribute to the Reich's formation by providing material welfare and ideal values.[101] Through his role as one of the founders of the Progressive Party (the party opposed to Otto von Bismarck), Virchow hoped to illuminate the way for

German science to guide the rest of German culture to the world's forefront. In an address entitled "Ueber die nationale Entwicklung und Bedeutung der Naturwissenschaften" ("On the National Development and Meaning of the Natural Sciences"), delivered at the fortieth meeting of the Versammlung Deutscher Naturforscher und Ärzte (Association of German Investigators of Nature and Physicians) in Hannover in September of 1865, Virchow recalled:

> As I entered into this lecture hall several days ago, I inspected the names of famous investigators of nature which our directors have written on the walls. The question rather vividly sprang forth from my soul, just how much our people have actually contributed to the development of the natural sciences, to the formation of a general human culture, and how much the increase of new names has guaranteed that we shall stand firm in our cultural-historical duty. . . . Perhaps you will say that this would be false pride, since it would not be the task of natural research to emphasize the national. It should be a cosmopolitan activity. We should all be parts, or members, of a large republic of scholars, and in this republic everything nationalistic should begin to vanish. It is precisely in this train of thought which these names excite in me that I feel justified in contradicting this view.[102]

Virchow continued by arguing what Lorenz Oken had proclaimed nearly 50 years earlier: ". . . there must be a German science, and that science must be set in the closest of relations with the life of the nation."[103] According to Virchow, science had greatly assisted in the moral liberation of the German people.[104] In 1871, immediately after the unification of the Reich, Virchow once again addressed the annual meeting of the Nationalversammlung Deutscher Naturforscher und Ärzte. His lecture was fittingly titled "Über die Aufgaben der Naturwissenschaften in dem neuen nationalen Leben Deutschlands" ("On the Tasks of the Natural Sciences in the New National Life of Germany").[105] Virchow had argued throughout the 1850s and the 1860s that the natural sciences simultaneously freed the Germans from the grasp of the Roman Catholic Church and helped forge a national identity to which all Germans could relate. Science, Virchow hoped, would alleviate the many forms of fragmentation that plagued the new Reich. Indeed, in reality, unified Germany was not as unified as many claimed.

As Cahan has argued, the Physikalisch-Technische Reichsanstalt epitomized the union of German scientific, technological, and entrepreneurial enterprises. Many Germans feared that their young nation lagged behind Britain and France in economic superiority, since the Reich did not support national institutions devoted exclusively to scientific research and precision technology as did their western rivals.[106] Hence, Siemens spearheaded the plans for a national institute, the PTR. This was, in Siemens's eyes, the most efficient way of becoming economically competitive. Helmholtz concurred: "It is obviously unworthy of a nation, which through its power and intelligence occupies and defends one of the foremost positions among civilized peoples, to relinquish to other nations or to the casual dilettantism of private, well-to-do individuals the care for the creation of fundamental knowledge."[107] The PTR, it was hoped, would end Germany's painful economic and national subservience. Reflecting upon Germany's dismal past achievements, Siemens lamented in his *Lebenserinnerungen*: "From earliest youth on, the disunion and powerlessness of the German nation pained me."[108] He claimed that the union of scientists with the technical business firms was necessary to ensure "the appropriate position in the great contest of the civilized world."[109]

The cultural and political roles of the German Handwerkerkultur were crucial to the invention of a German history of science throughout the nineteenth century. The fashioning of Fraunhofer as the orphan, working-class artist-genius had begun with Joseph von Utzschneider's biography, published a few months after Fraunhofer's death.[110] Fraunhofer and the solar lines were adopted into the canon of German physical sciences by Bunsen and Kirchhoff in the 1860s.[111] During priority disputes with British and French physicists, both of these men traced their scientific lineage to Fraunhofer. Helmholtz and Abbe, the other major contributors to the Fraunhoferian hagiography during the late nineteenth century, both wished to underscore Germany's recognition of the contributions of its artisans. Helmholtz and Abbe argued this in the 1880s, after Germany had become a Reich. Helmholtz claimed that the Handwerkerkultur was a uniquely Germanic phenomenon that was destined to make it a powerful nation in respect to France and England.[112]

The final touch in painting a portrait of cultural hegemony is the writing of a national history. This was precisely what German scientists were doing in the 1880s. These scientists used the genre of national histories to support both their young nation and their scientific acumen. Germany then possessed an impressive history, as France and Britain had centuries before.

The physicists Helmholtz and Abbe were not alone in the search for their historical roots. Virchow was writing a history of German biology. But the phenomenon of national-history writing was not confined to the sciences;

it was a much more ubiquitous cultural activity associated with German unification. Rudolf Haym's *Die romantische Schule*, published in 1870, offered a history of modern German literature tracing its origins back to the early Romantics. Earlier in the nineteenth century, musicologists, following the lead of Felix Mendelssohn-Bartholdy's Bach revival, were constructing a history of German musicology that depicted Bach as having begun a German tradition in music.

This chapter has focused on Helmholtz's and Abbe's historical accounts of Fraunhofer, German optics, and precision mechanics. Each physicist created a history reflecting his own interests. Helmholtz's account underscored Fraunhofer's humble origins and his ability, with the assistance of a reform-oriented government, to rise to the level of the Bürgertum, a class of Germans to which Helmholtz proudly belonged. Opticians and precision mechanics had contributed to Germany's success in the physical sciences. But with the decline of the status of skilled artisans, as a result of the industrialization of the early Kaiserreich, the future of German optics and precision mechanics was in doubt. Helmholtz's agenda was clear-cut. His historical account was, in part, aimed at the governmental officials present who were to vote on the approval of the PTR less than three weeks later. He wished to offer an account where science and technology worked hand in hand to raise revenue. He also wanted to point out that if instrument makers were recognized as a vital part of the scientific enterprise, Germany would be in a much better position to recapture the optical monopoly from its two rivals, France and Britain. For Helmholtz, the crafts tradition as exemplified by the members of the German Society for Mechanics and Optics was the key to success in precision optical technology.

Abbe's agenda was slightly different. He emphasized, to a much greater extent, Utzschneider's entrepreneurial skills and the patronage of King Maximilian I. These men transformed Bavaria into a first-rate, reform-minded kingdom where skilled artisans were rewarded. Abbe argued that Fraunhofer was the model for the German people. Fraunhofer was to inspire young German opticians and instrument makers serving the Reich. More important, Abbe hoped that the Prussian government would continue to emulate Bavarian patronage of Fraunhofer, for he and Schott received subsidies for their research on optical glass and apochromatic microscopy from the Prussian government. Abbe purposely depicted a portion of Utzschneider's enterprise that closely resembled what he himself was doing 50 years later. For Abbe, managerial practice that coordinated and preserved artisanal labor was just as important as the labor practices themselves for the establishment of precision optics in Germany. He wished to accomplish for Germany what Utzschneider had achieved for Bavaria—the coordination of science and technology—in order to reap financial success. Both Utzschneider and Abbe were concerned with the education and welfare of their skilled artisans. Abbe united optical theory with Schott's chemical manipulations in order to produce the world's finest microscopes and telescopes for the Carl Zeiss Works. Utzschneider united Guinand and Fraunhofer with Reichenbach to run the finest optical institute of its time. Abbe used history to support his enterprise and the work of skilled German artisans affiliated with precision mechanics.

Helmholtz's and Abbe's historical orations contained crucial similarities. First, both constructed Britain and France as the rivals of the Reich, economically, politically, and scientifically. Germany's envy resulting from its centuries-long inferiority to those two nations in the sciences in particular, and culture in general, was still apparent. But by the 1860s, Germany had begun to blossom in the sciences, whereas literature and music had already become national sources of pride generations earlier. Germany now needed a history to legitimize its new status in respect to France and Britain. Second, the importance of Guinand to the Bavarian optical glass enterprise was either ignored, as in Helmholtz's account, or underplayed, as in Abbe's. Moritz von Rohr, optician and optical historian, convincingly argued from the end of World War I until the 1930s that Guinand played a major role in the Optical Institute's production of achromatic lenses.[113] Interestingly, his historical accounts emphasized that Guinand was from Neuchâtel, which was ruled by Prussia during Guinand's lifetime.

In one very real sense, Helmholtz and Abbe's strategies worked. Abbe and Schott's lenses of barium borosilicate glass revolutionized the optical industry in the 1880s.[114] Abbe was able to offer superior lenses for both telescopes and microscopes. Germany's proclamation of optical supremacy was celebrated on the eve of a national anniversary of the German victory over the French at the Battle of Sedan. Elise Abbe, Ernst's wife, remarked:

> It was a beautiful September evening when the new glassworks were lit. . . . Bonfires were burning on the hillsides in preparation for the celebration, and bells were ringing. We were in such a happy, hopeful mood, that we forgot Sedan; the fires seemed to burn only for us, and the bells gave a proper consecration to our new beginning.[115]

Stuart Feffer argues that "the time was symbolic." "This," he continues,

> was to be a "glassworks for Germany" . . . and it too could be seen, in part, as an indirect victory of the German military. Much of the political backing for the subsidies that made the project possible had come either from the military directly or from suppliers of military goods and services. These subsidies were intimately tied to German ambitions in the national and international regulation of weights and measures—an initiative which itself had military support.[116]

The Chance Brothers and Feil began to lose their clients to the Carl Zeiss Works. A London customer, James Swift, reported the news of Abbe and Schott's new lenses and painted "a very gloomy picture as to the future of [Chance Brothers'] optical glass."[117] This gloomy outlook would loom over Britain's and France's optical enterprises until after the Second World War.

9 Conclusion

This study has offered a history of technological optics by focusing on optical glassmaking and lens production. Because Fraunhofer was a skilled artisan, his style of research was different from the mainstream practices of European optical savants; therefore, experimental natural philosophers in Britain and on the Continent were at a loss to explain just why he was so successful at manufacturing achromatic glass, lenses, and prisms. And, just as important, German physicists during the Kaiserreich could attribute to Fraunhofer many various characteristics that they claimed were essential to the formation of a precision optical technology. Fraunhofer became the rallying cry for many different causes. This work has discussed both how Fraunhofer constructed his optical glass and how his work was constructed by interested parties throughout the nineteenth century.

Helmholtz's depiction of Fraunhofer at the centenary celebration of the optician's birth was an attempt to get instrument makers included in the Physikalisch-Technische Reichsanstalt, whereas the goal of Abbe's account of Fraunhofer was a resumption of funding for his and Otto Schott's efforts to manufacture achromatic lenses for microscopes and telescopes. Abbe also wanted to encourage Prussian officials to support Germany instrument makers. With subsidies from the national government, it was hoped, Germany would regain an optical monopoly. The point to their speeches was historical: the newly created German Reich was destined to become a world power if it, unlike France or Britain, realized the importance of skilled artisans and the management of their labor to science, technology, economy, and therefore, the German nation. In late-nineteenth-century Germany, science was a revered cultural activity, and contextual histories unmask those accounts that depict the scientific enterprise with such awe.

The social practices that both generated and perpetuated those hagiographies inform the socio-cultural historian of the politics of disciplinary and national histories. The importance of skills and practices relevant to science and technology, and ultimately to the Reich, was a central issue. As most skilled workers were being proletarianized, instrument makers were being converted to trained technicians and were permitted less and less input into the development of scientific research. Abbe battled to give instrument makers more power in the decision-making process of the PTR. Relegation to quality control was both demeaning and unproductive. Hence, his Fraunhofer narrative underscored the scientific achievements of Fraunhofer and the entrepreneurial spirit of Utzschneider. In Abbe's view, the PTR should follow the example that he, Schott, and Zeiss were setting, and the Prussian Ministry should subsidize such work, as King Maximilian I once had.

Hagiographies are interesting not only because they highlight certain aspects of their subjects but also because they omit other aspects. Chapter 8 dealt with topics underscored by Helmholtz and Abbe: the importance of artisanal labor and the fusion of science, technology, and industry for the benefit of the state. But chapter 3 discussed crucial components of Fraunhofer's enterprise that neither scientist mentioned. First, Fraunhofer's laboratory was a secularized Benedictine monastery, and many of his assistants were monks. Second, because much of Fraunhofer's glassmaking technique was private knowledge, it stood in sharp contrast to what many claimed to be the character of scientific knowledge. Hence, this aspect, too, did not appear in either Abbe's or Helmholtz's celebration of Fraunhofer.

John Herschel, as was argued in chapter 5, claimed that artisanal skills and knowledge could be replicated, on a universal scale, by the rational principles of optics possessed by experimental natural philosophers such as he. Herschel never seemed to appreciate the skilled, practical knowledge necessary to produce optical glass. For him, the mathematics dealing with the curvature of the lens and the focal lengths, in addition to geometrical optics, was important in trying to reproduce Fraunhofer's workmanship. In his view, Fraunhofer's success was based upon the Bavarian's mastery of using the dark lines of the solar spectrum as a calibration technique for testing and manufacturing achromatic lenses. Like Babbage, Herschel wanted artisans to open their craft secrecy so that experimental natural philosophers could render that knowledge universally applicable to manufacturing.

Herschel and Babbage quintessentially belonged to the rational or scientific tradition of enlightened science. David Brewster disagreed, arguing that the labor of an artisan was his own property and that it should be protected by patent laws. Secrecy was the defense of skilled mechanics against piracy of their practical knowledge. Fraunhofer served Brewster's purposes perfectly, as the Scotsman attacked Britain's patent laws, taxation of crown and flint glass, and lack of scientific patronage. According to Brewster, Fraunhofer's success was due in part to the support of the Bavarian king. For Brewster, as for Faraday, the British lacked the necessary skills and practices to produce superior optical glass. At first Faraday believed that he could reverse engineer Fraunhofer's lenses and prisms by ascertaining the ingredients of the final product and their percentages. After years of failure despite the support of both the Royal Society and Board of Longitude, he decided to work on heavy glass composed of the silicated borate of lead, rather than flint, realizing that Fraunhofer's and Guinand's "private knowledge" might not be recoverable. Just how different Herschel's, Babbage's, Brewster's, and Faraday's responses were illustrates just how little they appreciated the complexities of precision optical technology, craft skills, secrecy, and managerial techniques.

In contrast with German accounts of the late nineteenth century, no one in Britain argued that the entrepreneurial tradition was of fundamental importance in understanding Fraunhofer's success, although the British held various views on the management of skilled labor. And it was the British, not the Bavarians, who emphasized the rational or scientific tradition in the 1820s and the 1830s. The Bavarians clung to secrets and craft traditions, as did the French in their enthusiasm for buying trade secrets. During the Kaiserreich, however, the Germans combined the craft tradition with the managerial and scientific traditions, never realizing how crucial the tradition of secrecy was to Fraunhofer's enterprise.

Similar issues still resonate today. Are specifiable rules sufficient for the construction of certain scientific artifacts (such as achromatic lenses)?[1] Which forms of the "practical knowledge" of scientists, engineers, and technicians can be replaced by machines? What is the relationship between artisanal and scientific knowledge? How should such types of knowledge be commercially managed? What are the social and epistemological differences between engineers and scientists? Although this book surely cannot answer those questions, it does show that one needs a historical perspective to begin to comprehend and explain the various answers offered today. And as was the case in the 1820s and the 1830s, answers to these undeniably complex questions are intricately and inextricably bound to larger socio-economic, cultural, and political issues.

My study of Fraunhofer has illustrated how differing responses to precision-technological practice in the nineteenth century were deeply embedded in more encompassing cultural beliefs. Whether one considered Fraunhofer's amalgamation of the secret and craft traditions to be science depended, in part, on one's views about his social status. Questions concerning the management of such labor depended crucially on whether that labor was communicable and, if so, how. The implications of these questions were (and still are) tremendous: How should scientific knowledge be taught to future scientists and technicians? How can technological, commercial firms achieve the critical balance between public and private knowledge to ensure market success and future viability? Answers to these questions not only shaped the society of pre-Victorian England and early-nineteenth-century Bavaria; they also affected the discipline of physics. In the nineteenth century, a professional class of (mostly) men began to emerge—the scientists—who dedicated themselves to the study of nature. As specialization of labor began to increase in society in general and within the scientific enterprise in particular, issues involving the nature and status of artisanal knowledge in respect to scientific knowledge and its management became more and more relevant—and indeed politically charged. The politics of labor can offer insights into how these issues were solved then and into how those solutions affect decisions being made today.

Appendix

Although there is no evidence that Fraunhofer actually derived an equation for calculating the refractive indices using prisms, one can very easily do so by using basic knowledge of optics, trigonometry, and Snel's Law. Fraunhofer, as an optician, undoubtedly possessed such knowledge, and standard eighteenth- and nineteenth-century textbooks on optical theory and practice assumed such knowledge. It also corresponds with the type of diagrams found in Fraunhofer's calculations of refractive and dispersive indices using symmetric passage.[1]

From Snel's Law applied to the entry face (AB) and the emergence face (CB) of the prism ABC, one has, with reference to figure A.1,

$$\frac{\sin\sigma}{\sin\beta} = \frac{\sin\rho}{\sin(\psi - \beta)}.$$

Simple mathematics yields

$$\sin\beta = \frac{\sin\sigma \sin\psi}{\sqrt{(\sin\sigma \cos\psi + \sin\rho)^2 + (\sin\sigma \sin\psi)^2}}.$$

Substituting back into Snel's Law, one obtains for the refractive index (n_{glass}) of the prism

$$n_{\text{glass}} = \frac{\sqrt{(\sin\sigma \cos\psi + \sin\rho)^2 + (\sin\sigma \sin\psi)^2}}{\sin\psi}.$$

In Fraunhofer's experiment the incident ray strikes at the same angle at which the emergent ray leaves (i.e., $\sigma = \rho$). In this particular case of symmetrical passage through the prism we find (with reference to figure A.2) the following:

$\mu \equiv$ the angle that the incident ray forms with the emergent, refracted ray.

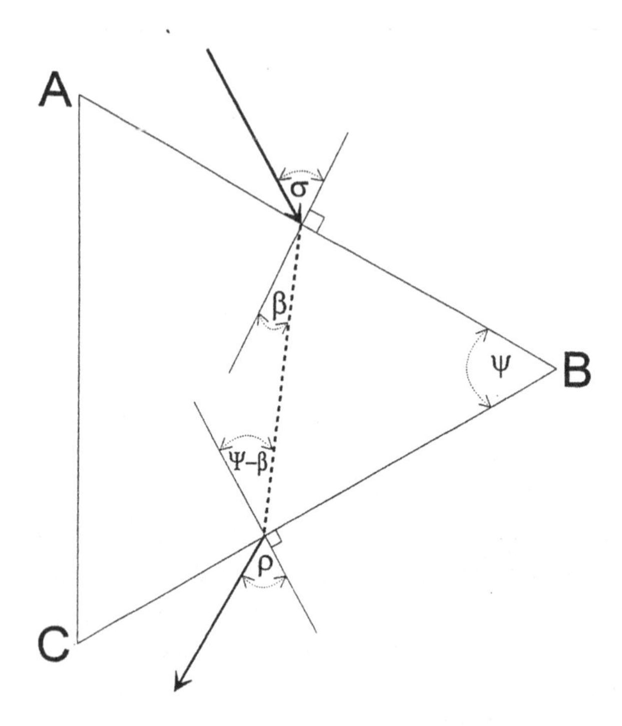

Figure A.1

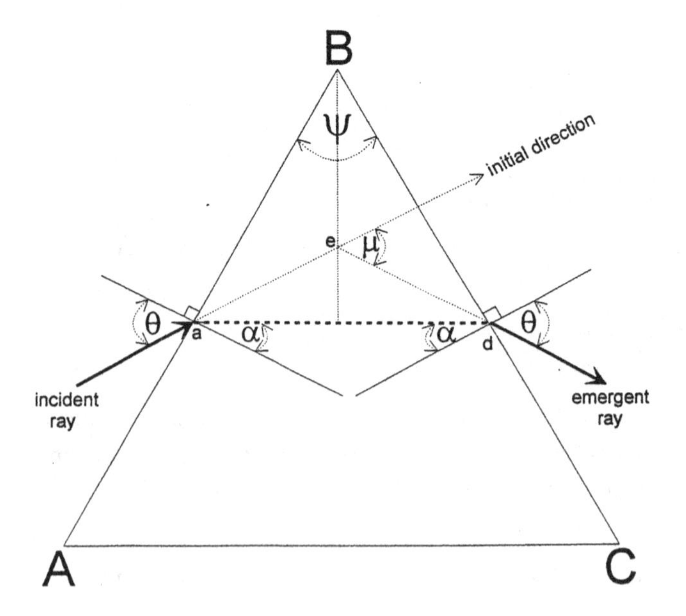

Figure A.2

$\alpha = \psi/2$, from symmetrical passage. The deviation angle μ is the sum of the two opposite interior angles in the triangle aed, or $\mu = 2(\theta - \alpha)$, whence

$$\theta = (\mu + \psi)/2.$$

Since, at point a, θ is the angle of incidence, and α is the angle of refraction,

$$\sin\theta = n_{glass}\sin\alpha,$$

where n_{glass} is the index of refraction of the glass prism with respect to the air. Thus, we have

$$n_{glass} = \frac{\sin[(\mu + \psi)/2]}{\sin(\psi/2)}.$$

Time Line

1787	Fraunhofer was born in Straubing, Bavaria.
1791	R. Blair's essay on liquid lenses was published.
1799	Fraunhofer began his apprenticeship with P. A. Weichselberger.
1801	Napoleon ordered the Bureau topographique to provide accurate topographical maps of Bavaria.
1802	W. H. Wollaston discovered the seven dark lines of the solar spectrum.
1803	Bavarian monasteries and cloisters were secularized.
1804	The Mathematisch-mechanische Institut was founded.
1805	J. Utzschneider purchased the cloister at Benediktbeuern and hired P. L. Guinand to produce achromatic lenses.
1806	A blockade severed Britain from the Continent.
	Utzschneider hired Fraunhofer to work at the Optical Institute.
1807	Fraunhofer wrote his first scholarly essay, dealing with catoptrics.
1809	Utzschneider asked Fraunofer to concentrate on dioptrics and to work with Guinand on production of optical glass.
1813	Guinand left Benediktbeuern after several quarrels with Fraunhofer.
	D. Brewster's *Treatise on New Philosophical Instruments* was published.
1817	Fraunhofer's essay "Determination of the Refractive and Dispersive Indices of Differing Types of Glass in Relation to the Perfection of Achromatic Telescopes" (written in 1814–15) was published.
	Fraunhofer was named a corresponding member of the Royal Bavarian Academy of Sciences.

1821 J. Herschel's essay "On the Alterations of Compound Lenses and Object-Glasses" was published.

E. Reynier contacted the Council of the Astronomical Society of London to sell P. L. Guinand's glass samples.

Fraunhofer was named an extraordinary visiting member of the Royal Bavarian Academy of Sciences.

D. Brewster co-founded the Edinburgh School of Arts (with L. Horner).

1823 H. Coddinton's *Elementary Treatise on Optics* was published.

1824 The Joint Committee of the Board of Longitude and the Royal Society for the Improvement of Glass for Optical Purposes was created.

J. Herschel visited Fraunhofer at Benediktbeuern.

1825 The Subcommittee for the Improvement of Glass for Optical Purposes was created.

Fraunhofer completed his Dorpat refractor.

The British offered Fraunhofer 25,000 pounds sterling for his secrets of producing optical glass.

Pellatt and Green began their experiments on the production of flint and crown glass.

1826 Fraunhofer died.

1827 The Board of Longitude contacted P. L. Guinand's relatives with regard to purchasing the secrets of manufacturing optical glass.

M. Faraday began his experiments on flint and crown glass at the Royal Institution.

1828 H. Guinand worked on optical glass with G. Bontemps of Choisy-le-Roi.

P. Barlow's liquid-lens telescope was built in Woolwich.

M. Faraday began working on heavy glass composed of silicated borate of lead.

The Board of Longitude approached G. Bontemps for the purchase of the secrets of manufacturing optical glass.

1829 J. Herschel resigned from the subcommittee.

Some of M. Faraday's heavy-glass disks were ground and polished into lenses by G. Dollond and inserted into telescopes.

H. Coddinton's *Treatise on the Reflexion and Refraction of Light* was published.

1830 D. Brewster's first assault on British patent laws was published.

The Royal Society formed the Telescope Committee, which tested M. Faraday's lenses.

Faraday effectively discontinued his research on optical glass.

Faraday's "On the Manufacture of Glass for Optical Purposes" was published.

1831 D. Brewster's *Treatise on Optics* was published.

1838 H. Guinand received the Lalande Medal for the communication of his father's secrets of producing flint glass.

1840 The Pulkovo refractor was built by G. Merz and J. Mahler.

1843 The lenses for the Harvard College refractor were completed by Merz and Mahler. Mahler died later in the year.

H. Guinand entered into agreement with his grandson, C. Feil.

1848 G. Bontemps left France during the Revolution and accepted a position at Chance Brothers in Birmingham, England.

1850 D. Brewster attacked the British patent laws in his Presidential Address to the BAAS.

1851 London's Great Exhibition illustrated French and British co-dominance in optical-glass manufacture.

1881 O. Schott and E. Abbe began their research on optical glass at the Carl Zeiss Company in Jena.

1887 Celebrations took place throughout Germany celebrating the hundredth anniversary of Fraunhofer's birth.

Abbreviations Used in Notes

BAW: JF	Bayerische Akademie der Wissenschaften (Munich). Akten der königlichen Akademie der Wissenschaften. Personal Akten: Herr Joseph Fraunhofer, 1821–26
BayHstA	Bayerische Hauptstaatsarchiv (Munich)
CUL.RGO	Cambridge University Library/The Board of Longitude Archives of the Royal Greenwich Observatory Manuscripts
DMA	Deutsches Museum (Munich) Archiv
RS.HS	Royal Society of London Archives, Herschel Letters
RS.CMB	Royal Society of London Archives, Committee Minutes Book
RS.DM	Royal Society of London Archives, Domestic Manuscripts
RS.MS.365	Royal Society of London Archives, Michael Faraday's Glass Furnace Notebook of the Royal Institution: October 1827–December 1829
RS.MM	Royal Society of London Archives, Miscellaneous Manuscripts
RS.MC	Royal Society of London Archives, Miscellaneous Correspondence
SPKB.Haus I.HA.NS	Staatsbibliothek Preußischer Kulturbesitz Berlin, Haus I (Unter den Linden, formerly Staatsbibliothek Ost Berlin), Handschriftenabteilung, Nachlaß Schumacher
SPKB.Haus II	Staatsbibliothek Preußischer Kulturbesitz Berlin, Haus II (Potsdamer Platz, formerly Staatsbibliothek West Berlin), Handschriftenabteilung

Notes

Chapter 1

1. William Hyde Wollaston had observed seven dark lines with a flint prism back in 1802, five of which he suggested could be used to mark precise regions of the solar spectrum. "A" bordered the red side of the spectrum, "B" separated the red from green, "C" was the limit of green and blue, while "D" and "E" demarcated the limits of violet. The lines "f" and "g" were not used for marking, as they could easily be confused with the "C" line. See Wollaston 1802, pp. 378–380. Fraunhofer was unaware of Wollaston's research until John Herschel informed him of it in 1824.

2. Rohr 1929a, pp. 4–14; Roth 1976, pp. 6–11; Sang 1987, pp. 9–11.

3. This is about 88 Bavarian florins (or guilders)—about half a year's salary for an apprentice's assistant in the Optical Institute during the late 1810s and the early 1820s, two months' salary for the glassmaker Pierre Louis Guinand of the Optical Institute in 1807, or half a month's salary for an ordinary professor at the University of Munich.

4. Such whiggish historiographies include Sutton 1976, Sutton 1972, and McGucken 1969. (See esp. pp. 4–10 in the last.) As James (1985b) has correctly argued, such historiographies have falsely asserted that the emergence of spectroscopy was a singular, continuous process of elaboration stretching back to Fraunhofer and extending to Gustav Kirchhoff and Robert Bunsen's establishment of the relationship between absorption and emission lines. See also Shapiro 1993, pp. 330–361.

5. A selected bibliography of Fraunhofer includes Bunsen and Kirchhoff 1860, Helmholtz 1887, Bauernfeind 1887, Jörg 1859, Jolly 1865, Seitz 1926, Brachner and Seeberger 1976, Merz 1865, and Abbe 1989. The twentieth-century biographies of Fraunhofer include Rohr 1929a, Roth 1976, Sang 1989, and Preyß 1989.

6. For example, Adam Smith argued in 1776 that the division of labor was the basis for the separation between mental and manual labor, and therefore between natural philosopher and laborer. See Smith 1976, vol. I., p. 201.

7. For example, the number of instrument makers who were elected into the Royal Society of London decreased dramatically by the 1830s as compared to the number that had been elected in the middle of the eighteenth century. See Bennett 1985.

8. Hobsbawm and Ranger 1983.

9. Reichenbach 1938; Popper 1959.

10. Popper 1959, pp. 45, 99. Interestingly, as Callebaut (1993, p. 207) has argued, Reichenbach and Popper's distinction between context of discovery and context of justification was taken from John Herschel's views of replication (discussed at length in chapter 5). See also Nickles 1980a,b.

11. Kuhn 1977, pp. 320–339.

12. These scholars, quite famously, have argued that logical positivism is invalid since there is no theory-independent language of observation. See Hesse 1963, Feyerabend 1962, Kuhn 1962, Hanson 1961, and chapter 2 of Holton 1978.

13. Latour 1983, pp. 141–170.

14. Gooding 1982; Pinch 1985, p. 14.

15. Galison 1987, 1997; Pickering 1991, 1995; Collins 1992; Shapin and Schaffer 1985; Schaffer 1989, 1988; Shapin 1994; Star 1991; Lawrence 1985; Latour 1990; Collins 1990; Anderson 1992.

16. Wittgenstein 1953, pp. 81–82, paragraphs 199–202.

17. Polanyi 1958, p. 53.

18. Ibid.

19. Ravetz 1996, pp., 75–108, esp. 78.

20. Collins 1992, p. 56.

21. Pinch, Collins, and Carbone 1996.

22. Olesko 1993, p. 16.

23. Ravetz 1996, p. 88.

24. See Buchwald 1993, pp. 169–209.

25. Bennett 1980, pp. 33–48; 1986, pp. 1–28.

26. Shapin 1994, pp. 355–407.

27. Secord 1994a, pp. 269–315; 1994b, pp. 383–408.

28. Sibum 1995, pp. 73–106.

29. Schaffer 1997, pp. 456–483; Morus 1998.

30. These four traditions are not transcultural, transtemporal categories; they are categories discussed by the historical actors themselves, which therefore can be used as tools by the sociocultural historian. Clearly there are overlaps between them.

Chapter 2

1. During the eighteenth century, works on optics could generally be divided into two categories: physical and geometrical optics. Physical optics deals with the nature of light, while geometrical optics is the mathematical theory of light that does not take light's physical properties into account. By the last quarter of the eighteenth century, a new field emerged: technical optics, or the design and construction of

optical instruments. See Atzema 1993, pp. vii and 1–3. Fraunhofer made significant contributions to physical optics (with his work on the diffraction gratings supporting the wave theory of light), to geometrical optics (with his work on dioptrics), and most famously to technical optics (with his improvement of the manufacture achromatic-lens).

2. King 1955, p. 144; Daumas 1989, p. 153.

3. Dollond 1753–54, p. 290.

4. Ibid.

5. Ibid., p. 291.

6. King 1955, p. 146; Riekher 1990, p. 107.

7. Dollond 1757–58, p. 735.

8. Ibid., p. 736.

9. Ibid., p. 737.

10. Ibid., p. 739.

11. Ibid., p. 740.

12. Ibid., p. 741.

13. Riekher 1990, p. 112.

14. Klingenstierna 1759–60, pp. 944–77; King 1955, p. 156.

15. Clairaut hoped to provide a complete theory of achromatics by tying in Dollond's work with his earlier essays. See Clairaut 1756, 1757, 1758.

16. King 1955, p. 156.

17. Ziggelaar 1993, p. 365.

18. Boegehold 1935, p. 97.

19. Gilbert 1810, p. 245.

20. Gilbert 1810, p. 247; Boegehold 1935, pp. 99–106.

21. King 1955, p. 157.

22. Dollond 1765, pp. 55–56. Fraunhofer accomplished both of these feats with only two lenses.

23. Ziggelaar 1993, p. 364.

24. Ibid.

25. Boscovich 1765.

26. Ziggelaar 1993, pp. 364–365.

27. Ibid., p. 366; Boscovich 1785.

28. King 1955, p. 158. The secondary spectrum was nearly entirely eradicated by Fraunhofer during the second decade of the nineteenth century; it was finally completely corrected by Ernst Abbe and Otto Schott of the Carl Zeiss Works in the 1880s.

29. Brewster 1813, p. 354.

30. These papers were subsequently published: Blair 1794, pp. 3–76.

31. Ibid., 43.

32. Lenses of any kind can be formed from liquid media by adding the fluid between glasses, which have one side formed as convex and the other concave with the same radius. The glass itself must have a very small refractive power.

33. Blair 1794, pp. 43–44.

34. Ibid., p. 45.

35. Smith 1738, vol. 2, p. 281.

36. Ibid., pp. 312–317 and 301–312.

37. Smith 1778.

38. Brewster 1813, pp. 273–274.

39. Ibid., p. 274.

40. Ibid., p. 275.

41. Ibid., p. 279.

42. Ibid., pp. 297–298.

43. Ibid., p. 298.

44. Ibid.

45. Ibid., pp. 303–306.

46. Ibid., pp. 389–396.

47. Ibid., 397–398.

48. Ibid., p. 399.

49. For a detailed account of Wollaston's device, see Buchwald 1985, pp. 19–23.

50. *The Artisan* 1825, pp. 164–165.

51. Brewster 1813, pp. 243–246.

52. Herschel 1821, pp. 222–267.

53. Ibid., p. 222.

54. Brewster 1832, p. 607.

55. Herschel 1821, pp. 222–223. "D'Alembert has proposed, among others, to annihilate the spherical aberration for rays of all colours, a refinement which might almost be termed puerile, were it not for the respect due to so great a man." (ibid., p. 226)

56. Herschel 1821, pp. 226–227.

57. Ibid., pp. 223–224.

58. Ibid., p. 250.

59. Ibid., pp. 252–253.

60. Brewster (1827b, p. 6), looking back at how Fraunhofer's technique revolutionized optical precision, commented: "Hitherto, achromatic object-glasses had only been calculated for rays proceeding from a point in the axis of the lens, but Fraunhofer considered the deviations for all points situated without the axis, and this is always a minimum in his object-glass."

61. Herschel 1821, p. 252.

62. Herschel 1822a, p. 362.

63. Ibid.

64. Ibid.

65. Ibid., pp. 363–364.

66. Ibid. p. 364.

67. Ibid., pp. 364–365.

68. As quoted in King 1955, pp. 194–195 (originally appearing in the article "Telescope" in Rees's *Cyclopaedia* of 1819).

69. This joint committee is discussed at length in chaper 6.

70. Royal Society of London Archives, Herschel Letters (henceforth RS.HS.), 26.45 folio 1.

71. Herschel 1822b, esp. pp. 244–245. This paper was read on 7 May 1821 before the Philosophical Society of Cambridge.

72. Herschel 1823.

73. Herschel 1822b, p. 446; my emphasis.

74. Ibid., p. 449.

75. Ibid., pp. 458–459.

76. Ibid. p. 458.

77. Ibid. p. 459.

78. The phrase "irrationality of the coloured spaces" refers to the fact that the colored spaces do not have the same ratio to one another as the lengths of the spectra which they compose.

79. Herschel 1822b, p. 460.

80. Herschel 1822a.

81. Ibid., p. 363 (my emphasis). Ideally, these colors are red and blue or violet, but they depend upon the medium used. As Herschel remarked, the brightest rays do not always correspond to the extreme rays of the spectrum (ibid., p. 364).

82. Ibid., p. 364.

83. Barlow 1827a.

84. Ibid., p. 231.

85. Ibid., p. 232.

86. Barlow 1828, p. 111.

87. Airy 1827, p. 241. The paper was written in 1824.

88. Coddington 1823.

89. Ibid., p. iii.

90. Ibid., p. 93.

91. Ibid.

92. Ibid., p. 94.

93. I am using Coddington's variables. Here dispersive power is Δm, whereas for Herschel's equation above dispersive power was $\partial\mu/(\mu - 1)$. Also, here p designates a radius, whereas Herschel used p to denote dispersive power.

94. Coddington 1823, p. 96. This is the more accurate version of John Dollond's 3:2 ratio, determined some 60 years earlier.

95. Coddington 1829.

96. Ibid., p. i. In his preface, Coddington discusses how optical theory had led to new advances that took hold between the first and second editions with regard to telescope construction. He mentions G. B. Airy's work on achromaticism and spherical aberration of eyepieces published in the *Cambridge Philosophical Transactions*, and Robison's work on the telescope published in the supplement to the *Encyclopaedia Britannica*. He later discussed Fraunhofer's contributions.

97. Coddington 1829, pp. 237–238.

98. Ibid., p. 239.

99. Ibid. p. 240.

100. Ibid. p. 251.

101. Coddington draws upon Herschel's truncated series for determining the first and second order of the dispersive power of a substance as outlined in Herschel 1821, pp. 250–253.

102. Brewster 1831, pp. 87–88.

103. Ibid., p. 68.

Chapter 3

1. The Optical Institute was originally located in Munich as a part of the Mathematisch-mechanische Institute from 1804 until the end of 1806. It returned to Munich in 1819, although the production of optical glass remained in Benediktbeuern under Fraunhofer's close supervision.

2. The most comprehensive accounts of the skilled labor of monks in glassmaking are Lerner 1981, Vopelius 1895, Haller 1975, Friedl 1973, Lobmeyr 1874, and Blau 1954 and 1956.

3. For a discussion of these texts, see Ganzenmüller 1956, pp. 46–51. Ganzenmüller lists Theophilus's *De Diversis Artibus* of 1123 (it should be noted that Theophilus was a Bavarian Benedictine monk), Paracelsus's references to glass and the divine light, and Oswald Croll's *Basilica Chymica* of 1609. Johannes Kunckel's seventeenth-century work *Ars Vitraria Experimentalis* discusses texts dealing with glass and the divine light written during the sixteenth and seventeenth centuries.

4. Utzschneider himself even conducted achromatic glass trials in the Benedictine Cloister of Ettal's glasshouse in Grafenanschau, but those trials failed (Kirmeier and Treml 1991, p. 342).

5. Fraunhofer, "Ueber die Construktion des so eben vollendenten grossen Refraktors," in Lommel 1888, pp. 169–170. For an English translation of this essay, see *Philosophical Magazine and Journal* 66 (1825): 41–47.

6. See Klügel 1778. Fraunhofer used this text extensively while manufacturing achromatic lenses.

7. Sheehan 1989, pp. 225–226.

8. On the financial crisis of Bavaria at the turn of the eighteenth century, see Utzschneider 1837.

9. Archives du Ministère des Affaires Étrangères, Paris: Corr. politique Bavière 178, Alquiers Report to Talleyrand, 6 Ventrôse an VII, 24 February 1799, reprinted in Weis 1983b (p. 565).

10. Weis 1983b, pp. 559–560; Weis 1971. See also Walch 1935, p. 9.

11. Weis 1983b, pp. 571–572.

12. Ibid., pp. 559–595; Nipperdey 1991, pp. 11–101.

13. Bayerische Hauptstaatsarchiv (henceforth BayHstA), Abteilung IV, Kriegsarchiv, A VI, 2 Bd. 2.

14. Although the official order came on 19 June 1801, archives show that the project was being planned nearly a year before. See ibid. and Deutsches Museum Archiv (henceforth DMA), Joseph von Utzschneider Nachlaß, 5220, 5221, 5251, 5254, 5257, 5259, 5260, 5263, 5265, 5268, and 5271. Compare Amann 1908, pp. 1–2; Messerschmidt 1976, p. 11.

15. DMA 5251. See also 5263, 5265, 5268, 5271.

16. DMA 5254.

17. BayHstA IV, A VI, 2 Bd. 2; Amann 1908, p. 2.

18. Schiegg later became a professor at Würzburg. He received an annual salary of 2000 guilders as scientific advisor to the Bureau topographique. See BayHstA IV, A VI, 2 Bd. 2

19. Messerschmidt 1976, p. 11.

20. Utzschneider 1826, p. 7.

21. BayHstA IV, A VI, 2 Bd. 2, 17 September 1805.

22. Seitz 1923, pp. 150–154; 1927a, pp. 134–135.

23. Bauernfeind 1885, p. 8. See also Soldner's letter to King Maximilian I, 20 January 1807: BayHstA IV, A VI, 2 Bd. 2

24. Soldner 1911.

25. Ibid., p. 1.

26. See, e.g., Niedersächsische Vermessungs- und Katasterverwaltung 1955; Großmann 1955a, pp. 45–54; Großmann 1955b, pp. 371–384; Gerardy 1955, pp. 54–62.

27. Veit 1962.

28. Bauernfeind 1885, pp. 9, 14.

29. Copie de la lettre de la Direction du Bureau topographique de Bavière à M. le Chef de Brigade Bonne, 10 Mars 1808, BayHstA IV, A VI 2 Bd. 2.

30. Prince Elector Maximilian Joseph IV became King Maximilian I of Bavaria in 1806.

31. BayHstA MFS 56638; Anordnung 8 September 1808, BayHstA IV A VI 2 Bd. 2.

32. BayHstA IV, A VI 2 Bd. 2, Anordnung von 30 April 1801.

33. Ibid. As a result of Soldner's triangulation calculations, Utzschneider appointed him advisor to the Steuerkatasterkommission on 13 March 1811. See also Bauernfeind 1885, p. 8.

34. Amann 1920, p. 3.

35. Ibid., pp. 3–7; Bauernfeind 1885, p. 5. See also Benzenberg 1818.

36. Amann 1920, p. 6

37. Bauernfeind 1885, p. 17; Roth 1976, pp. 49-50; Dyck 1912, p. 41; originally reported in 1812 by Staatsrat Klüber, curator of the Mannheim Observatory.

38. The Royal Academy of Sciences in Munich donated 600 guilders to the project; the Franco-Bavarian Bureau topographique donated 100 guilders (Roth 1976, pp. 36–38).

39. Dyck.1912, pp. 13–14.

40. Reichenbach 1821.

41. Dyck 1912, pp. 18–19.

42. Ibid., pp. 9 and 23.

43. DMA 6045.

44. Utzschneider 1826, pp. 4–5.

45. Dyck 1912, p. 19.

46. Financially, the Optical Institute was far less successful. From 1806 to 1813, Utzschneider had spent over 60,000 guilders on optical glass research. Montgelas provided some state subsidy, to the tune of 70,000 guilders. On 2 March 1818 Utzschneider sold Kloster Benediktbeuern back to the state for 250,000 guilders. See Roth 1976, pp. 64–65 and 94.

47. Rohr 1926, pp. 121–123. Utzschneider searched for his glassmaker in monasteries because they were the major sites of glass production throughout that region for nearly 1000 years.

48. The manuscript is on display at the Deutsches Museum in Munich. A German translation can be found in Rohr 1928a.

49. For the details of Utzschneider's travels and subsequent hiring of Guinand, see Seitz 1926, pp. 13–23. See also DMA 7276, 7279, 7281, 7282, 7284, 7285, and 7289.

50. Chance 1937, p. 434; Seitz 1927b, p. 314; Seitz 1926, pp. 22–23. For the notes on Guinand's and Utzschneider's optical glass trials, see Staatsbibliothek Preußischer Kulturbesitz Berlin. Haus II. Handschriftenabteilung. Sammlung Darmstaedter (henceforth SPKB.Haus II.HA.SD.), Fraunhofer Nachlass. Box 4: 1–7, 37, 41–44, 81–82.

51. Utzschneider had experience in converting secularized monasteries into workshops. In 1802 he had founded a leather and soap factory in the former Carmelite monastery. Eventually employing 170 workers, it became one of Munich's largest industrial concerns. See Kirmeier and Treml 1991, p. 341.

52. Weis 1991, p. 28. See also Weis 1983a.

53. Weis (1991, p. 28) claims, quite correctly, that this was very much a Germanic phenomenon.

54. Weis 1991, p. 28.

55. Chadwick 1975, p. 122.

56. Weber 1991. The *Carmina Burana*, a collection of light verse from the twelfth and thirteenth centuries, was found in the library of Benediktbeuern.

57. BayHstA, Kl. Benediktbeuern 118, folios 7–23.

58. Weber 1991, p. 58.

59. See Imhof 1806. See also Koch (n.d.), pp. 12–13.

60. Stutzer 1986, pp. 137–146.

61. Ibid., p. 175.

62. Ibid., pp. 173–175.

63. As quoted in Weis 1991, p. 31.

64. Ibid.

65. Weis 1983a, pp. 45–46.

66. The official recommendation is reprinted in Weis 1983a (pp. 68–77).

67. Bauer 1991, p. 36.

68. Utzschneider 1826, p. 10.

69. Ibid., pp. 8–9.

70. Fraunhofer, "Über parabolische Spiegel und Beschreibung krummliniger Segmente, in Anwendung auf die Verfertigung elliptischer-, parabolischer- und hyperbolischer Spiegel zu Theleskopen," SPKB.Haus II.HA.SD.Fraunhofer Nachlass, Box 2: 13–26 and 631–646. This essay, lost during the publication of Lommel's collection of Fraunhofer's essays, resurfaced early in the twentieth century in Berlin and was subsequently published by Rohr (1929b, pp. 42–52). It is dated March 1807.

71. See Utzschneider 1826, p. 11.

72. The details of Fraunhofer's polishing machine and the description of the techniques for polishing are found in the SPKB.Haus II.HA.SD.Fraunhofer Nachlass, Box 5. This information was never published, as it was considered a company secret. See Sang 1987, p. 40. It is clear that British opticians were decades behind in this optical technology. Indeed, they were still producing inferior parabolic lenses in 1827. See Cecil 1827, esp. pp. 6–99.

73. Riekher 1990, pp. 151–152. In the Fraunhofer archives (SPKB.Haus II.HA.SD.Fraunhofer Nachlass. Box 2: 183–188) there are notes from an article by Klügel published in 1796 in the *Abhandlungen der Göttinger Gesellschaft der Naturwissenschaften* on the refraction of of rays through a double object lens. The notes are dated 1797 and therefore predate Fraunhofer's interest in optics. They are not written in his hand, but he most likely consulted them. There are also notes on Klügel's *Analytische Dioptrik*. Fraunhofer undoubtedly drew upon Klügel's 1810 essay "Angabe eines möglichst vollkommnen achromatischen Doppel-Objectivs."

74. Klügel 1778, pp. 57–63 and 96; Rohr 1929b, pp. 43–44.

75. Rohr 1929b, pp. 43–44.

76. SPKB.Haus II.HA.SD.Utzschneider Nachlass, Box 1: 8.

77. SPKB.Haus II.HA.SD.Utzschneider Nachlass, Box 1: 14 (6 August 1809).

78. DMA 6079; Utzschneider 1826, p. 11.

79. Utzschneider 1826, p. 11. One notice from 1820, "Das Handbuch für Reisende in der südlichen Gebirge von München," claimed that approximately 100 workers were employed by Utzschneider at Benediktbeuern, most of whom worked for Fraunhofer (Seitz 1926, p. 37).

80. Utzschneider 1826, pp. 18–19. See also Howard-Duff 1987, p. 341.

81. Utzschneider 1826, p. 18.

82. SPKB.Haus II.HA.SD, Utzschneider Nachlaß, Utzschneider and Fraunhofer correspondence.

83. Utzschneider 1826, p. 19.

84. Ibid., p. 18. From 1814 (the year of his third contract) to 1819, Fraunhofer's monthly income was 125 guilders, plus profits earned, which typically added 700 guilders a year, plus living quarters and wood for heat. He also received an astonishing 10,000 guilders in 1814, since by that time he shared in the Optical Institute's ownership with Utzschneider. Fraunhofer was then set for life. From 1819 until his death in 1826, Fraunhofer earned 150 guilders per month, plus approximately 700–800 guilders per year from profits. In 1823, after being named Konservator of the Royal Academy of Sciences' Mathematical and Physical Instrument Collection, he received an additional 800 guilders per year. (See Utzschneider 1826, p. 19.) It should be noted that 10,000 guilders is just less than 6 times the annual salary that Lorenz Oken received as an ordinary professor of biology at the University of Munich in 1819. (See DMA 7323; Sang 1987, p. 73; Rohr 1929a, p. 196; Seitz 1926, pp. 90–91.) His assistants were generally earning between 18 and 24 guilders per month in 1820 (SPKB.Haus II.HA.SD.Fraunhofer Nachlaß, Box 5, "Lohne der Arbeiter").

85. Indeed, Fraunhofer carefully controlled what he taught and to whom. In a letter (SPKB.Haus II.HA.SD.Utzschneider Nachlass. Box 1: 59, 28 July 1810), Utzschneider tells Fraunhofer that the unskilled lads from the community of Reichenhall "must only remain a certain time, three to six months, so that they do not become aware of the work in the Institute."

86. I unearthed these archives in SPKB.Haus II.HA.SD, Fraunhofer Nachlass, Box 4. There are scores of recipes, dated by Fraunhofer himself. An example of the ratios of the ingredients: 181 parts (by weight) of red lead, 154 parts of quartz, 47 parts potash, 9 parts saltpeter ("Flint Glass # 18").

87. Ibid.

88. Ibid. One recipe called for 83 parts sand to 65 parts potash to 27 parts calcium carbonate.

89. The ovens at Benediktbeuern could reach 1300°C, whereas coal furnaces could reach 1500°C.

90. SPKB.Haus II.HA.SD.Fraunhofer Nachlass. Box 4, "Bemerkungen während des Schmelzens im grossen Ofen."

91. Guinand had noted in 1805 that bubbles could be avoided if a high heat was applied to the molten glass for ten to twelve hours. See Rohr 1928a, pp. 450–451. For Guinand and Fraunhofer, the key was to stir thoroughly, yet gently, as not to produce bubbles and not destroy the glass' transparency (ibid.). For a characterization of the mechanical device, see Jebsen-Marwedel 1987, pp. 17–18. For reasons of secrecy, Fraunhofer never detailed this procedure as Faraday did (see chapter 6 of this volume; see also Faraday 1830). The details of Guinand's and Fraunhofer's techniques are reconstructed as much as possible from the few notes that Fraunhofer left behind ("Besondere Bemerkungen während des Schmelzens im grossen Ofen" and "Bemerkungen während des Schmelzens im kleinen alten Ofen" in SPKB.Haus II.HA.SD.Fraunhofer Nachlass. Box 4: Blue Folder III, Numbers 1–7) and from Guinand's notes dating back to 1805 (SPKB.Haus II.HA.SD.Fraunhofer Nachlass. Box 4: 1–44) as well as his original manuscript on the preparation of optical glass. (See Rohr 1928a.)

92. Lommel 1888, pp. 3–31. The essay originally appeared in *Denkschriften der königlichen Akademie der Wissenschaften zu München für die Jahre 1814 u. 1815*, Bd. V (not published until 1817). Fraunhofer sent this paper to his good friend Johann von Soldner, mathematician, ordnance surveyor, and member of the Royal Academy. Inclusion in this periodical meant instant recognition as a Naturwissenschaftler, something that the working-class artisan desperately sought.

93. Lommel 1888, p. 3.

94. Ibid., p. 27.

95. Ibid.

96. David Brewster and John Herschel also attempted, in vain, to isolate monochromatic light. See Brewster 1823.

97. Although Fraunhofer never described his lamp, it was most likely a burning wick drawing salted alcohol.

98. There can be overlap of colors from the intermittent lamps; i.e., the lamp adjacent to B might supply prism H with a small portion of violet rays as well. However, since prisms map angle onto position, and since the incident rays fall parallel (as a result of such a large distance between the two prisms) on prism H, all lamplight of the same wavelength will be mapped onto some position as seen by the telescope of the theodolite.

99. See, e.g., SPKB.Haus II.SD.Fraunhofer Nachlass, Box 2: 445.

100. Symmetric passage with respect to the D line.

101. Lommel 1888, p. 10.

102. As I stated earlier, Wollaston had been the first to realize that the solar dark lines could be employed to demarcate precise positions of the spectrum back in 1802, but Fraunhofer was unaware of Wollaston's work until 1824.

103. Lommel 1888, pp. 13–14.

104. See, e.g., Boscovich 1767, p. 142; Klügel 1810b, pp. 276–291.

105. See the appendix to this volume.

106. It is this part of Fraunhofer's scientific and technological enterprise that is very analogous to cooking. Many master chefs argue that it is difficult to describe, particularly in writing, the attributes that inform the trained eye that their creation is a success, such as its smell, taste, or texture. Similarly, in glassmaking, knowing when and how long to stir, and when the glass is cool enough to manipulate, separates the master from the rookie apprentice. Indeed, sometimes apprentices never learn such skills. I would like to thank Julia Child for her helpful conversations on the necessary skills and practices for cooking and how they are communicated.

107. Since Fraunhofer never published this information, he and Utzschenider must have felt that some portion of this type of knowledge could indeed be communicated in algorithmic fashion.

108. See, e.g., volume 2 (1816): 165–172.

109. Prechtl 1828, p. 178. This list was published after Fraunhofer's death, but price lists were undoubtedly published in other optical texts from about 1818 onward.

110. SPKB.Haus II.HA.SD.Utzschneider Nachlass, Box 1: 183–185. Letters to Fraunhofer: 15 and 16 June and 9 July 1814. This contradicts MacKenthun's claim (1958, p. 131) that the czar did indeed visit the institute, although the point is really moot.

111. SPKB.Haus II.HA.SD.Utzschneider Nachlaß, Box 1: 37.

112. As quoted in Seitz 1926, p. 23.

113. As quoted in ibid., pp. 49–50.

114. Fraunhofer's letter to Schweigger, 2 August 1817, DMA 7414.

115. RH.HS.2.199, Herschel to Babbage, 3 October 1824.

116. As quoted in Eamon 1994, p. 34.

117. I shall employ Henri Lefebvre's (1984) notion of social space as a heuristic tool in order to offer a microhistorical account of how Fraunhofer manufactured his lenses.

118. Weber 1991, p. 58.

119. Kirmeier and Treml 1991, p. 232; SPKB.Haus II, Fraunhofer Nachlaß, box 5, "Lohne der Arbeiter."

120. Schmitz 1982, p. 148.

121. "Aufzeichungen des Lehrers Aloys Rockinger in Benediktbeuern," reprinted in Seitz 1926, p. 95.

122. Stutzer 1986.

123. Kirmeier and Treml 1991, p. 342.

124. Ibid., pp. 260–61.

125. Fellöcker 1864.

126. Eamon 1994, p. 81. On guild ordinances restricting access to secrets during the early modern period, see Long 1991 and Shelby 1976.

127. Eamon 1994, p. 81.

128. Some glassmakers encoded their recipe books with numbers, each number corresponding to a particular letter. Those who could not break the code could not reproduce the glass. One example of number-coding was found in a recipe book from a monastic glasshouse in the forests of Bavaria. The key to the puzzle was decoding the word "crystal." The letters of the German word Christalen corresponded to the numbers from one to ten, from left to right, that is, c was 1, h was 2, r was 3, and so forth. See Blau 1954, pp. 51–56 and 175–178; Lerner 1981, pp. 46–47. It should also be noted that distilling and brewing processes, which were executed in Benediktbeuern from the sixteenth century until the secularization in 1803, were originally closely related to secret alchemical processes.

129. Eamon 1994, p. 7.

130. Ibid., p. 30. For a history of craft recipe books, see Merrifield 1849.

131. Smith 1994, p. 37.

132. Friedl 1973, pp. 11–17. This particular work discusses the importance of secrecy in glassmaking as reported in the various stories and myths of Bavarian glassmakers.

133. Gasquet 1966, p. xxiii.

134. Ibid., pp. xxvi, 84.

135. Ibid., pp. 6–97.

136. Workman 1913, pp. 219–220. See also Ovitt 1986, p. 88.

137. Weber 1976, pp. 118–119; Ovitt 1986, p. 89. I concur with Ovitt that Weber was misguided by claiming that capitalist origins lay in Benedictine culture. Capitalism views labor and its products quantitatively, as a means of increasing productivity. Monasticism cherishes the labor process itself. (See Ovitt 1986, p. 106.)

138. Mumford 1967, p. 264; Ovitt 1986, p. 89.

139. Mumford 1967, p. 264.

140. Ibid.

141. Ovitt 1986, p. 104.

142. On the importance of labor to the Benedictines, see White 1971.

143. Gasquet 1966, pp. 19–20.

144. Ibid., p. 77.

145. Ibid., p. 93.

146. Ibid., pp. 118–119.

147. Eschapasse 1963, pp. 11–21.

148. Ibid., p. 17.

149. These included glass huts, sawmills, slaughterhouses, brandy distilleries, and breweries. See, e.g., Kirmeier and Treml 1991, pp. 186–193.

150. "Idleness is an enemy of the soul. Because this is so the brethren ought to be occupied at specified times in manual labour, and at other fixed hours in holy reading. We therefore think that both these may be arranged for as follows: from Easter

to the first of October, on coming out from Prime, let the brethren labour till about the fourth hour. From the fourth till close upon the sixth hour let them employ themselves in reading. On rising from the table after the sixth hour let them rest on their beds in strict silence. . . . Let None be said somewhat before the time, about the middle of the eighth hour, and after this all shall work at what they have to do till evening." (Gasquet 1966, pp. 84–85)

151. Giddens 1984, esp. chapter 3; Hillier and Hauser 1984, pp. ix–xi, 4–5, 8–9, 19; Foucault 1980, pp. 63–77; Foucault 1979; Ophir 1984.

152. Both ways are suggested by Lefebvre (1984).

153. Ibid., pp. 32–36.

154. Ibid., 86–87.

155. Shapin 1988, pp. 373–404. Shapin's article can be seen as an elaboration of Owen Hannaway's work on the laboratory as a site of knowledge production (1986, pp. 585–610). Other works on the laboratory as a social space of knowledge production include Galison 1985 and Owens 1985.

156. Newman 1999.

Chapter 4

1. See, e.g., Jörg 1859, Jolly 1865, and the numerous works of Rohr. See also Helmholtz 1887 and volume 2 of Abbe 1989. For a somewhat detailed accounted of Fraunhofer's struggles for recognition, see Sang 1987.

2. According to Woodmansee (1994, pp. 29–30), the author Karl Philipp Moritz was considered to be a shameless recycler of his own ideas.

3. Woodmansee 1994, p. 36.

4. Jones 1975, p. 274.

5. Woodmansee 1994, p. 48.

6. As quoted in Woodmansee 1994 (p. 48) and in Pape 1970 (columns 103–104). See also Reich 1773.

7. Eisenlohr 1856, p. 11. See also Woodmansee 1994, p. 53.

8. Gieseke 1957, p. 122. See also Woodmansee 1994, p. 53.

9. Such logic also prevailed in seventeenth-century England. See Shapin 1994, pp. 355–407. Similarly, by the late eighteenth century and the early years of the nineteenth, the Royal Society of London would only accept instrument makers who specialized in one area of instrument making (e.g., optical instruments). They would not admit instrument makers who catered to a large market by not specializing. Society members thought such a move by artisans represented crass commercialism and was not worthy of academic respectability. I thank Jim Bennett his helpful discussions on this matter.

10. DMA, Fraunhofer Nachlaß, N14/6 5414 a/b. Fraunhofer was not the first Bavarian Handwerker to be admitted to the Royal Academy of Sciences in Munich. Some 60 years earlier, the renowned instrument maker Georg Friedrich Brander

(1713–1783) was honored with such a membership. However, Brander came from a middle-class household, as his father was a Kaufmann (merchant), and secrecy was not such a major issue for his work. Also, he studied mathematics and natural sciences at the University of Altdorf (Nuremberg). Hence, Fraunhofer was a rather different case in a rather different period. Fraunhofer, during the first two decades of the nineteenth century, was, to my knowledge, the only Handwerker in the Academy. In this respect Bavaria was similar to Britain, as instrument makers were much more readily recognized by academies in the eighteenth century than in the early nineteenth. On Brander, see Friedrich 1910.

11. Bayerische Akademie der Wissenschaften, Akten der königlichen Akademie der Wissenschaften, Personal Akten: Herr Joseph Fraunhofer, 1821–26 (henceforth BAW:JF), folio 1.

12. Ibid.

13. Ibid., folia 8–9. See also Sang 1987, pp. 96–97.

14. BAW:JF, folio 8.

15. Here 'Künstler' here means *artiste* and the nineteenth-century definition of *artist*, rather than its twentieth-century English meaning.

16. BAW:JF, folio 8.

17. Ibid.

18. Yelin 1818, pp. 601–608.

19. BAW:JF, folia 16–17.

20. Ibid., Akt V, 12, no. 25, 4 April 1820.

21. Ibid.

22. This might sound confusing to the modern reader. In German states, ordinary memberships and ordinary professorships were higher in stature and salary than extraordinary members and extraordinary professorships. Ibid., Wahlakt 1820, 27; 1821, folio 28.

23. Ibid., folio 28 (2 August 1821). See also Roth 1976, pp. 99–100.

24. BAW:JF Wahlakt 26; 31 July 1821.

25. DMA, Fraunhofer Nachlaß, N14/27, 5417.

26. See SPKB.Haus II, Fraunhofer Kasten 3, letter no. 7; and DMA, Fraunhofer Nachlass, 14/9, 5415; and 14/7, 5420 a/c.

27. Zschokke 1817, p. 559.

28. Ibid., p. 560.

Chapter 5

1. As quoted on p. 263 of Stewart 1998.

2. Dodsworth 1987, p. 8.

3. Howard 1940, p. 21.

4. By 1696, 60 percent of glass factories in England had been located in London, Bristol, Stourbridge, and Newcastle; while Worcester, Gloucester, and Nottingham could only claim a mere 5 percent. These data are taken from a letter from John Houghton, dated 15 May 1696, listing all the glasshouses in England and Wales (reprinted in Powell 1923, p. 39).

5. Powell 1923, p. 39.

6. Ironically, by the end of the eighteenth century, the amount of lead oxide added to flint glass was so great that it could no longer be made homogeneous.

7. Daumas 1989, p. 157.

8. Ibid., p. 156. Original source: Cassini 1810, p. 180.

9. Daumas 1989, p. 157.

10. Gilbert 1810b, pp. 460–461.

11. Gilbert 1810a, pp. 251–252.

12. Sorrenson 1993. The status of instrument makers would fall drastically in Britain by the 1810s and the 1820s. This will be a critical point in my story; it explains, in part, why Bavaria was to usurp England's optical reign at that time.

13. On the conflict between Dollond and Chester Moor-Hall, see Sorrenson 1993.

14. Daumas 1989, pp. 228–245, 258–291.

15. I thank the curators of the Göttingen Observatory and the Gotha Observatory Museum for permitting me to look at their instruments.

16. Bennett 1985, pp. 13–27.

17. Thompson 1963, p. 261.

18. Morrell and Thackray 1981, p. 10.

19. Thompson 1963, p. 269.

20. Powell 1923, pp. 156–157.

21. Douglas and Frank 1972, p. 31. This is also found in *Thirteenth Report of the Commissioners* (1835).

22. Dodsworth 1987, p. 13.

23. Porter 1832, p. 142. Again, this was repeated in *Thirteenth Report of the Commissioners* (1835).

24. Porter 1832, p. 141.

25. Ibid., pp. 142–143.

26. Ibid., p. 143.

27. Powell 1923, p. 153.

28. Pellatt 1849, pp. 67–68.

29. Guttery 1956, p. 86.

30. Powell 1923, p. 153.

31. *Thirteenth Report of the Commissioners* 1835; Powell 1923, p. 154; Guttery 1956, p. 86.

32. *Thirteenth Report of the Commissioners* 1835; Powell 1923, p. 155; Guttery 1956, pp. 87–88.

33. Guttery 1956, p. 89.

34. Matsumura 1983, p. 13.

35. Indeed, it was illegal to manufacture duty-free and taxed glass on the same premises (Guttery 1956, p. 90).

36. Matsumura 1983, pp. 14–15.

37. As quoted in ibid., p. 19. This comment certainly described the labor practices involved in glassmaking decades earlier. Although it was generally considered that "this extreme sub-division of employment [results in] the superior skill of workmen and the excellence of the manufacture" (*Artizan* 1 (1844), p. 233), it hampered the British subcommittee's attempt at trying to bring these practices together to produce Fraunhoferian-quality optical glass. See chapter 6. For a general discussion of science and technology's impact on the labor aristocracy, see Berg 1980, pp. 153–161.

38. As quoted in Matsumura 1983, p. 23.

39. As quoted in ibid.

40. Dodd 1843, p. 131.

41. Pellatt 1849, p. 68.

42. Douglas and Frank 1972, p. 32.

43. Ibid.

44. Faraday 1830, p. 2. See also Cambridge University Library/Board of Longitude Archives of the Royal Greenwich Observatory Manuscripts (henceforth CUL.RGO) 14/10.167 and 168 (11 and 21 May 1825) and RS.HS.8.113.

45. Guttery 1952, p. 23.

46. For informative discussions on the changing notion of skill in early-nineteenth-century Britain, see Rule 1987, Prothero 1979, Behagg 1982, and Morus 1998.

47. See Morse 1981.

48. Shapiro 1993, pp. 330–361.

49. SBPK.Haus I, Handschriftenabteilung, Nachlaß Schumacher (henceforth SPKB.Haus I.HA.NS), letter to Schumacher from Brewster, 23 June 1821, folio 1. Fraunhofer thanked Schumacher for publicizing his work. See SPKB.Haus I.HA.NS, Fraunhofer Correspondence, 21 September 1821. For the details and specifications of the order see Fraunhofer Correspondence, 1 April, 22 July, and 8 October 1822, and 25 January 1825.

50. SPKB.Haus I.HA.NS, Brewster to Schumacher, 29 October 1821, folio 5.

51. Brewster 1823, 1824. See also, Shapiro 1993, pp. 331–354.

52. Recall that this paper was published in 1817.

53. Brewster 1832, pp. 678–681.

54. SPKB.Haus I.HA.NS, folio 7.

55. Ibid., Brewster to Schumacher, 21 December 1824, folia 9–10.

56. Ibid.

57. SPKB.Haus I.HA.NS, Brewster to Schumacher, 3 January 1825, folia 11–12.

58. SPKB.Haus I.HA.NS, Brewster to Schumacher, March (?) 1825 (the letter is undated but was received on 4 April), folio 17. For some reason, Schumacher did not acknowledge the order until 17 January 1826, when he wrote to Fraunhofer describing the order and congratulating him that his reputation had now spread to England. SBPK.Haus II, Fraunhofer Nachlaß, box 5, Schumacher to Fraunhofer, 17 January 1826.

59. *Edinburgh Journal of Science* 2 (1825), p. 174.

60. Brewster 1825a, pp. 305–306. Recall that flint glass 4 inches in diameter was considered the limit due to homoegeneity.

61. Brewster 1826b, p. 110.

62. Brewster 1825b, p. 348.

63. Brewster 1827a, pp. 10–11. This quote was reprinted in Babbage and Brewster 1830.

64. SPKB.Haus I.HA.NS, Brewster to Schumacher, 24 November 1826, folia 13–14.

65. Ibid.

66. SPKB.Haus I.HA.NS, Utzschneider Correspondence, 20 June 1826; new schedule for fulfilling delayed orders, 20 and 25 October 1826; 29 December 1826; and 16 January 1827.

67. SPKB.Haus I.HA.NS, Brewster to Schumacher, 16 December 1827, folio 20.

68. Henry Fox Talbot to John Herschel, RS.HS.17.261, July 1826.

69. Henry Fox Talbot to John Herschel, RS.HS.17.265, 18 July 1827.

70. *Edinburgh Review* 60 (1835), p. 392.

71. For the classic study of this debate, see Morrell and Thackray 1981.

72. Brewster 1821, p. 114.

73. Ibid.

74. Ibid., p. 115.

75. Ibid.

76. Ibid.

77. Shapin 1984, p. 20. See also Berg 1980, pp. 158–159.

78. Morrell 1974, p. 40.

79. Shapin 1984, p. 20.

80. Ibid.

81. *The Scotsman*, No. 294, 7 September 1822, p. 283.

82. Brewster 1824b, pp. 203–205.

83. Ibid., p. 203.

84. Ibid., p. 204.

85. Ibid.

86. Ibid.

87. Brewster 1827a, p. 11.

88. For an excellent discussion of the individualist/heroic interpretation versus a social deterministic interpretation of invention during the nineteenth century, see MacLeod 1996. See also Coulter 1992.

89. During the 1850s, when Brewster returned to the theme of Britain's patent laws, he dismissed those who wished to abolish completely the patent laws as practicing what he referred to the "wildest socialism." See Brewster 1855, p. 263. As MacLeod has argued (1996, p. 150), Hodgskin exercised considerable influence on Karl Marx.

90. Babbage and Brewster 1830, pp. 305–342. Babbage is credited with writing the first part of the essay dealing with England's scientific decline; Brewster authored the part of the article on British patent laws (pp. 333–342).

91. Ibid., p. 333.

92. Ibid., p. 337. It is interesting to note that Brewster does know of examples where workmen reading the specification could not reproduce the item to be patented.

93. Ibid.

94. Gordon 1869, pp. 208–209. This was a reprint of a portion of Brewster's presidential address (Report of the British Association for the Advancement of Science 1850 (London, 1851), pp. xxxi–xliv).

95. Brewster 1855, p. 261.

96. Ibid., p. 249.

97. Ibid., p. 261.

98. Babbage and Brewster 1830, p. 338.

99. Ibid., p. 340. Other examples of inventors denied their just rewards included Samuel Crompton, inventor of the spinning mule, who died impoverished in 1827, and Richard Trevithick, a leader of design for high-pressure steam locomotives, who died equally poor in 1833. See MacLeod 1996, p. 137.

100. Brewster 1855, p. 210.

101. Ibid., pp. 258–262.

102. Morrison-Low 1984, p. 60.

103. As quoted in ibid., p. 61.

104. Gordon 1869, p. 95.

105. As quoted in Morrison-Low 1984, p. 61. For the original source, see the correspondence and papers of James David Forbes (1809–1868), held in the University Library of St. Andrew's University.

106. Ferguson 1823, volume 1, p. vi.

107. Ibid.

108. Brewster 1845, p. 190.

109. Ibid., pp. 190–191.

110. Brewster 1825b, pp. 353–354.

111. Brewster to Faraday, 8 August 1832 (James 1991–1997, volume 2, p. 74, letter 605).

112. Ibid.

113. Brewster 1845, p. 192.

114. Ibid.

115. Brewster 1826a, p. 283.

116. Ibid., p. 284.

117. Blair 1791, pp. 1–76.

118. For a summary of Blair's research, see Brewster 1832, p. 612.

119. Ibid.

120. Brewster 1826a, p. 285. As will be discussed in the next chapter, the subcommittee did investigate the possibility of substituting flint glass with liquid lenses filled with the sulfuret of carbon.

121. Evans 1981.

122. Shapiro 1993, pp. 331–354.

123. Guinand to Herschel, 20 April 1822, RS.HS.13.134.

124. Dollond, Herschel, and Pearson 1826; Rohr 1924, p. 792.

125. Dollond, Herschel and Pearson (1826, p. 511) noted that Tulley needed to spend much time and effort to produce a piece of crown glass to use with Guinand's flint glass. Gilbert, Herschel, and Pearson concluded that "no degree of excellence in individual specimens would authorise them to recommend their purchase by the Society, unless supported by such assurances of constant supply, as would render it a matter of public interest." Nothing ever came of it.

126. As will be discussed in the next chapter, this committee was created in response to Guinand's and Fraunhofer's superior optical lenses and prisms. Britain was no longer the world's purveyors of optical glass. The Joint Committee wanted to invent a process that would yield optical glass of consistently Fraunhoferian quality.

127. Herschel's diary entries relevant to Fraunhofer are reprinted in Rohr 1928b.

128. Herschel to Babbage, 3 October 1824, RS.HS.2.199.

129. Ibid. See also Seitz 1926, pp. 111–112; *Edinburgh Journal of Science* 2 (1825), pp. 344–348.

130. Herschel to Babbage, 3 October 1824, RS.HS.2.199.

131. Herschel to Littrow, 9 November 1824, RS.HS.11.248.

132. Fraunhofer gave a glass prism to Henry Fox Talbot, who passed it on to Herschel in London in 1825 (DMA, Fraunhofer Nachlaß N14/20, 5389). George Dollond had previously requested that Herschel bring back glass samples from Munich (RS.HS.6.498 (31 March 1824)).

133. Letter 332, Herschel to Faraday, Royal Institution Manuscripts F3 B397 (James 1991–1997, volume 1, p. 438).

134. It turns out that, although that particular prism was superior, Merz and Mahler, Fraunhofer's successors, generally could not match Fraunhofer's craftsmanship. See chapter 7 below.

135. RS.HS.26.45, folio 1. See note 70 to chapter 2.

136. Seitz 1926, p. 112.

137. Herschel to the Reverend Brinkley (Andrews Professor of Astronomy at Trinity College, Dublin), RS.HS.4.277. No date is given for the letter, but it must have been written between 3 October and 9 November 1824.

138. Ibid.

139. Babbage 1830, pp. 210–211.

140. Herschel to Babbage, 3 October 1824, RS.HS.2.199; Seitz 1926, pp. 111–112; *Edinburgh Journal of Science* 2 (1825), p. 348. On the notion that observation is a skill, see Hacking 1983, p. 180.

141. Robert Fot to Herschel, 20 February 1840, RS.HS.7.348.

142. Prothero 1979, p. 198.

143. E.W., "London Mechanics' Institute," in *The Mechanics Weekly Journal* (London, 1824), pp. 76–77. Originally in No. V, 13 December 1823.

144. Herschel 1830, p. 15.

145. Ibid., pp. 69–70 (emphasis added).

146. Ibid., p. 63.

147. Ibid., pp. 70–72.

148. DMA, Fraunhofer Nachlaß, N14/20, 5389; Herschel 1826, p. 293.

149. Herschel's Minutes of the General Scientific Committee, 1827, RS.HS.26.38.

150. Herschel 1830, p. 352.

151. Ibid., p. 354.

152. King 1955, pp. 120–243. See also Chapman 1996, section XIII, pp. 5–8. British hegemony in regard to the astronomical reflector was never questioned.

153. King 1955, p. 206.

154. Talbot to Herschel, 27 February 1826, RS.HS.17.259.

155. Struve 1828, p. 82. See also Struve 1825.

156. Struve 1826a, p. 97. See also Struve 1826b, pp. 443–455.

157. Fraunhofer, "Ueber die Construktion des so eben vollendenten grossen Refraktors," in Lommel 1888. For an English translation of this essay, see "On the Construction of the Large Refracting Telescope just completed by M. Fraunhofer," *Philosophical Magazine and Journal* 66 (1825): 41–47.

158. *Philosophical Magazine and Journal* 66 (1825), p. 43. The classification of epsilon Boötes was the trial most often used as an illustration of the superiority of refracting telescopes. Friedrich Wilhelm Bessel, the professor of astronomy and

director of the observatory at Königsberg, used a Fraunhofer refractor to classify epsilon Boötes as a double star of the first class, not of the fourth class, as William Herschel had claimed earlier (ibid., p. 42).

159. Herschel 1826, p. 289.

160. Ibid.

161. Ibid., p. 290.

162. Ibid., p. 293. On the protocol for gentlemanly disputes, see Shapin 1994.

163. South 1826, p. 287.

164. King 1955, p. 206.

165. See, e.g., Alder 1997, pp. 62, 135; Schaffer n.d.

166. Schaffer 1994, pp. 203–227.

167. Giedion 1969, pp. 35–36.

168. Schaffer 1994, p. 214

169. Ibid.

170. Ashworth 1996, p. 631.

171. Babbage and Brewster 1830, p. 341.

172. Williams 1981.

173. Brande was clerk at the Royal Mint, professor of chemistry at the Royal Institution, editor of the Royal Institution's journal, and foreign secretary to the Royal Society; he also served as chemist to the East-India Company.

174. Faraday 1830, p. 10.

175. Greenaway et al. 1971–1976, 1: 1. See also Cantor, Gooding, and James 1991, pp. 27–31.

176. Cantor, Gooding, and James 1991, p. 29; Gooding and James 1985.

177. Cantor, Gooding, and James 1991, p. 31. Faraday did his utmost to educate the public in science. Indeed, the 1820s witnessed the rise of a genre of journals aimed at educating working-class artisans in the theories of science, including optics: "By endeavoring to give a popular form to the various subjects that come under their notice, the Conductors of the "Artisan" . . . may, in some degree, advance the interest of science, by promoting its general diffusion, at a time when there is so much anxiety shown by every class of the community to acquire a knowledge of the accurate sciences." (*Artisan* (1825), p. 2; see also James 1992)

178. Faraday 1827, p. iii. Faraday clearly considered Herschel the philosopher of their enterprise. Indeed he wrote to Herschel on 10 November 1832, "When your book on the study of Nat. Phil. came out I read it . . . with delight. I took it as a school book for philosophers and I feel that it has made me a better reasoner & even experimenter and has altogether heightened my character and made me if I may be permitted to say so a better philosopher." (RS.HS.7.179)

179. Faraday 1827, pp. iii and vi.

180. Ibid., pp. vi–vii.

181. Ibid., p. vii.

182. Ibid.

183. Faraday 1830, p. 4.

184. Ibid.

185. Ibid., pp. 1–2.

186. Ibid., p. 2.

187. Rule 1987, pp. 104–105.

188. Cooper 1835, pp. v–vi.

189. Ibid., pp. vi–viii.

Chapter 6

1. *Edinburgh Journal of Science* 2 (1825), p. 348.

2. Brewster 1831b, p. 144.

3. Brewster 1831a, p. 67.

4. Ibid., pp. 67–68.

5. Ibid., p. 359.

6. Simms 1852, pp. 7–8.

7. CUL.RGO.14/8.17–20. James (1991a, esp. pp. 30–31; 1991b, esp. pp. 37–38) has argued that the Royal Society subcommittee was created to bail out the financially doomed Board of Longitude. I shall take another line, as the reader will see. I think that both James's point and mine need to be considered as joint reasons for the subcommittee's creation. They certainly are not mutually exclusive.

8. CUL.RGO.14/8.20. This was an act of Parliament.

9. Royal Society of London, Committee Minutes Book (henceforth, RS.CMB).1, 127; Royal Society Domestic Manuscripts (henceforth, RS.DM) volume 3, folio 26; and RS.DM.3.22.

10. See, e.g., RS.CMB.1.96 and 101.

11. RS.CMB.1.96–97. Apsley Pellatt Jr. was one of Britain's leading glass manufacturers. It was often the case in the artisanal world in general—and glassmaking in particular—that fathers would work with their sons and/or sons-in-law. It must be reemphasized that British opticians, such as George Dollond, did not produce their own glass, but purchased glass banks from glassmakers such as Pellatt and Green. This stood in sharp contrast to Fraunhofer.

12. CUL.RGO.14/10.159. See also 14/8.22. In July, Pellatt and Green informed Young that the approximate cost of the furnace would be more expensive than originally anticipated. The new estimate was £500. They were also preparing clay pots to hold the molten glass. By the time the entire furnace was erected, the cost was £878.19.4. The Joint Committee agreed to pay £700, the remainder to be paid for by Pellatt and Green. CUL.RGO.14/10.161 and 163; and 14/8.31.

13. Pellatt to Young, CUL.RGO.14/10.161.

14. Pellatt to Young, CUL.RGO.14/10.165.

15. George Harrison to Commissioner of the Excise, 11 May 1825, CUL.RGO.14/10.167. See also the letter of the Treasury Chambers to Davies Gilbert dated 21 May 1825, CUL.RGO.14/10.168.

16. RS.CMB.1.101.

17. RS.MS.364.1, Journal of the Experiments for the Production of Glass for Optical Purposes, 1825–26; RS.CMB.1.100.

18. RS.MS.364.1.

19. RS.CMB.1.132.

20. RS.CMB.1.100–2.

21. Faraday 1830, p. 3.

22. RS.HS.26.45, folio 1. See also RS.HS.26.47.

23. RS.CMB.1.126–27; RS.MS.365.95–96.

24. Faraday (1830, p. 8) listed some extreme examples where the specific gravity of samples from the top of the crucible were significantly lower than samples from the bottom: 3.73 versus 4.63, 3.85 versus 4.74, and 3.81 versus 4.75.

25. RS.HS.7.166.

26. RS.CMB.1.120–1.

27. RS.HS.26.38, folio 2; RS.MS.364, 10–11.

28. Herschel to Gilbert, 17 April 1827, HS.26.30.

29. Ibid.

30. Herschel's Minutes of the General Scientific Committee of 1827, RS.HS.26.38; and RS.CMB.1.147.

31. Herschel's Minutes of General Scientific Committee of 1827, RS.HS.26.38.

32. RS.CMB.1.157. Herschel wrote to Gilbert that Pellatt and Green could not manufacture the desired quality glass blanks (RS.HS.26.30). He suggested that a regular operator be employed in order to carry out the daily experiments that Faraday, Dollond, and he would suggest.

33. RS.CMB.1.157–58, 174–79; Royal Society Miscellaneous Manuscripts (henceforth RS.MM.)14.3. The hearth of the oven was 41 by 22 by 18 inches. See also RS.MS.365.2–3.

34. Faraday 1830, p. 3. Even Herschel seemed impressed by Anderson, calling him "[in] all respects suitable and likely so far as they can judge to prove useful not merely as a labourer, but purposes of a higher class" (Herschel's Committee Report of 1827, RS.HS.26.37).

35. Herschel to Faraday, 6 November 1827, RS.HS.26.34.

36. Ibid.

37. RS.CMB.1.147–49.

38. Michael Faraday's Glass Furnace Notebook of the Royal Institution: October 1827–December 1829 (hereafter RS.MS.365), pp. 6–7.

39. Ibid., pp. 14–15.

40. RS.CMB.1.100. One of those lenses was manufactured by Pierre Louis Guinand.

41. RS.MS.365.16–23.

42. Herschel to Faraday, 26 October 1828, RS.HS.26.41.

43. RS.CMB.1.131–35, 177–78.

44. RS.MS.365.199–200.

45. RS.MS.365.200.

46. RS:MS.365.227–8.

47. During the early nineteenth century, 'platinum' referred only to the metal, which constitutes the greatest part of the mineral platina. 'Platina' had two different senses. The first—the one intended by Faraday—was the crude ore of platinum, a type of metallic sand from South America that forms a very strong metallic substance. It contains platinum, palladium, iridium, osmium, and rhodium. The second definition—a regional one, used in the British Midlands, particularly by Birmingham metal workers and button manufacturers—was a pale brass in which the red color of copper is so diluted by a large proportion of zinc that it appears nearly white. It is usually referred to as Birmingham platina. See *Mechanics Weekly Journal*, No. XII, 31 January 1824, p. 191.

48. Faraday 1830, pp. 10–13.

49. Ibid., pp. 20–32.

50. RS.CMB.1.196; RS.HS.26.38, folio 1.

51. RS.CMB.1.120–1 and 204. The specific gravities of Faraday's glass samples were nearly double those of Fraunhofer's.

52. Faraday 1830, p. 34.

53. Ibid.

54. Ibid., p. 35.

55. Ibid., p. 36.

56. Ibid., p. 37.

57. RS.MS.365.193.

58. Faraday 1830, pp. 37–38; RS.MS.365.357. Spongy platina is used platina dissolved in a mixture of strong hydrochloric and nitric acids and water. The muriate of ammonia is added to this solution, forming a bright yellow precipitate. The precipitate is washed several times with water, and allowed to dry on a filter, and then heated slowly until all the vapors are removed.

59. RS.MS.365.357–8.

60. RS.MS.365.360.

61. RS.MS.365.371.

62. RS.MS.365.424.

63. Faraday, Royal Institution Glass Furnace Notebook, volume 2 (RS.MM.XIV.1.3–7).

64. RS.MM.XIV.7.

65. Presumably, Faraday's glass samples were not good enough to produce the dark lines. See RS.CMB.1.130 and RS.HS.26.44.

66. RS.HS.26.41. This letter illustrating how glass made from the silicated borate of lead could potentially suffer striae to a greater degree than flint glass was dated 26 October 1828. Again, Herschel was interested in geometrical optics as a solution of Faraday's practical problem of manufacturing optical glass.

67. Herschel to Gilbert, 21 July 1828, RS.HS.8.113.

68. Herschel to Faraday, 21 July 1828, RS.HS.7.176.

69. Faraday 1830, p. 2.

70. RS.MS.365.264 and 360, for example. It should be noted that several of Faraday's glass blanks can be found at the Royal Science Museum in London.

71. RS.MS.365.274.

72. Faraday numbered this glass sample 130 (RS.MS.365.265–270).

73. RS.MS.365.361.

74. Ibid.

75. RS.DM.3.154.

76. Herschel to Gilbert, 19 March 1830, RS.HS.8.126.

77. Ibid.

78. RS.DM.3.155.

79. RS.DM.3.157.

80. RS.DM.3.159.

81. RS.DM.3.160.

82. As quoted in James 1991b (p. 39) and in James 1991–1997 (volume 1, letter 446).

83. Royal Society of London Miscellaneous Correspondence (henceforth RS.MC.) 1.316.

84. SPKB.Haus II.HA.SD.Fraunhofer Nachlass, Box 4, "Kleine Versuchen mit Glasschmelzen."

85. Faraday 1830, p. 15; Roth 1976, p. 55.

86. Rogers 1829, p. 229.

87. Littrow 1831, pp. 481–482.

88. King briefly discusses liquid lenses in his *History of the Telescope* (1955, pp. 189–191). See also Brewster 1827b, p. 335.

89. Barlow wrote: "My object is (as I wish distinctly to be understood) not to supplant the use of flint glass in the construction of this instrument, but to supply its place by a valuable substitute in cases where the former cannot be obtained sufficiently large, or where it can only be obtained at an expense which must always limit the possession of a good astronomical telescope to persons of fortune and to public institutions." See Barlow 1827b, p. 326.

90. Brewster 1827b, p. 335.

91. Ibid.

92. Blair 1827.

93. Barlow 1828a,b, 1829.

94. Barlow 1827b, p. 324.

95. Barlow 1832, pp. 353–357; Barlow 1828c; Barlow 1828a, p. 112.

96. Barlow 1832, p. 353; 1828c.

97. Barlow 1832, p. 354.

98. Ibid.

99. CUL.RGO.14/8.58. On the subcommittee's decision to investigate Blair's technology of liquid lenses, see RS.CMB.1.207.

100. Barlow 1832, p. 360.

101. *Proceedings of the Royal Society* 3 (1831–1837), pp. 245–253. Herschel viewed another liquid lens telescope contructed by Barlow in 1836 and concluded that the liquid had leaked out of the lens (letter, John Herschel to George Dollond, 31 October 1836, RS.MC.2.230).

102. Dollond et al. 1826, p. 507.

103. Ibid., p. 508.

104. Ibid., p. 509.

105. Herschel to H. Guinand, 9 October 1827, RS.HS.26.29.

106. Charles Tulley to Herschel, [25] October 1827, RS.HS.26.32. See also Tulley's report to the Astronomical Society on 25 January 1825: "M. Guinand died . . . about the end of 1823, and about the seventy-sixth year of his age, immediately after arrangements had been made with the French government for the purchase of his secret. His son fortunately possesses all of the details of the process, and is ready to supply opticians with glass for object-glasses of large apertures." (Brewster 1825b, p. 354)

107. Charles Tulley to Herschel, [25] October 1827, RS.HS.26.32.

108. Roth 1976, p. 120. This was originally reported by the director of the Royal Bavarian mint, Heinrich J. von Leprieur, when Fraunhofer gave him power of attorney. See also Rohr 1929a, p.154; Preyß 1989, p. 74. As Herschel informed those wishing to procure the secrets that Fraunhofer was bound by contract not to divulge any information concerning the production of optical glass, the attempt was indeed bribery.

109. RS.CMB.1.158–59; RS.DM.3.41; RS.HS.26.38.

110. In the original manuscript, M. Guinand's name was written, then crossed out, and Bontemps's name was added.

111. RS.CMB.1.211 and RS.DM.3.53

112. RS.DM.3.62–70.

113. RS.DM.3.63.

114. RS.DM.3.62.

115. This was reported by Roget to Bontemps on 31 December 1828 (RS.DM.3.67).

Chapter 7

1. "Eigene Worte des Akademikers und Opticus Dr. von Fraunhofer einige Wochen vor seinem Tode zum Nachschreiben gesprochen an seinen Freund von Leprieur," in Swimme 1924. See also Roth 1976, p. 52; Preyß 1989, p. 74.

2. BayHstA MK 41347 (Das optische Institut Utzschneider und Fraunenhofer [*sic*] in München und die Frauenhoferischen [*sic*] Original-Papiere im Kaiserlichen Hausarchiv betrachtet, 1826– , no. 597). See also BayHstA MInn 43981; DMA 7346; Seitz 1929, p. 55.

3. BayHstA MK 41347; BayHstA MInn 43981.

4. BayHstA MK 41347 (originally marked Hausarchiv 41/5), p. 11.

5. Ibid., no. 18927.

6. Ibid., no. 794.

7. As quoted in Seitz 1926, p. 101.

8. Pauli had been particularly praised by Fraunhofer.

9. Seitz 1929, p. 55; DMA 7436.

10. DMA 7404, 25 June 1826, Utzschneider to Merz.

11. DMA 7410.

12. "Glasversuche Utzschneiders in Benediktbeuern nach Fraunhofers Tod, 1827–1829," DMA 5357. Utzschneider had Fraunhofer's ingredients and their percentages of his flint and crown glass as well as the technical procedure for procuring those glass blanks in hand.

13. 19 March 1829, DMA 5357.

14. Stahl 1929, pp. 151–52; Seitz 1929, pp. 55–57.

15. Stahl 1929, p. 152.

16. Ibid., pp. 152–153.

17. They did receive an award for the best traditional astronomical instrument.

18. As quoted in Bennett 1983, p. 4. Original source: *Lectures on the Results of the Great Exhibition of 1851* (1852), p. 348.

19. Rohr 1933, pp. 414–445.

20. Bontemps 1868, pp. 652–653. Hence, the Royal Society of London heard that Lerebours had access to Guinand and Fraunhofer's secrets and commissioned him to manufacture glass samples. On Lerebours's workmanship, see *Edinburgh Journal of Science* 8 (1828), p. 363; 9 (1828), p. 170. Note also the following (Brewster 1825b, p. 353): "Among the opticians who have used [Pierre Louis Guinand's] glass may be mentioned M. Lerebours, a French artist, who, during a visit to Brenets in

1820, obtained all the glass which M. Guinand then had, and was so well satisfied with it that he requested a fresh supply, and made overtures for obtaining the process. We may also mention M. Cauchoix, who, in a notice relative to the telescopes in the last exhibition at the Louvre, has spoken highly of the flint-glass of which they are constructed."

21. Rohr 1933, p. 415.

22. Chance 1919, p. 172. Clearly, Bontemps' procedure was unreliable in 1828. As we saw in the previous chapter, one of his flint-glass disks provided to the English in that same year was of "inferior quality"; the other was better but "of little value." As time went on, he improved his technique.

23. Bontemps 1868, pp. 663–664; Chance 1919, p. 173.

24. Utzschneider 1830, p. 351; original source: *Astronomische Nachrichten,* no. 163 (1829). For Guinand's claim of priority of the invention of achromatic glass used in Benediktbeuern, see Dollond et al. 1826, esp. p. 507.

25. Utzschneider 1830, p. 351.

26. Ibid., pp. 352–353. See also Utzschneider 1829.

27. Letter from Herschel to Schumacher, 25 February 1825, SPKB, Haus II, Fraunhofer Kasten 5. E. Reynier was P. L. Guinand's apprentice in Les Brenets after Guinand left Benediktbeuern. See R[eynier] 1824.

28. Bontemps 1868, p. 653; Abbe 1989, volume 4, p. 79.

29. Rohr 1933, p. 416.

30. Chance 1919, p. 50.

31. Bontemps 1868, pp. 653–654; Chance 1919, pp. 5–6.

32. Chance 1919, pp. 10–11.

33. Ibid., pp. 6–7.

34. Ibid., pp. 52–53.

35. Bennett 1983, p. 4; Chance 1919, p. 176. See also Bontemps 1851.

36. Bennett 1983, pp. 4, 5.

37. Chance 1919, p. 54.

38. Ibid., p. 175.

39. Ibid., pp. 175–176.

40. Ibid., pp. 95, 178.

Chapter 8

1. Cahan 1989, p. 24.

2. Helmholtz 1887, p. 115. The original German reads:

Ein hoher Geist hat diesen Leib bewohnt,
Ein Geist, der jetzt in seiner Heimath thront.
Denn seine Heimath war die Erde nicht,

Die Sternenwelt war's und das ew'ge Licht.
Ein Adler, der sich auf zur Sonne schwang,
Und ihres Lichts Geheimnisse durchdrang.
Die Sterne folgten seinem mächt'gen Ruf,
Dem zaubervollen Glase, das er schuf.
Entrissen hat er sie der alten Nacht,
Getheilt, verdoppelt und uns nah gebracht.
Der hinter Sternen ruh'nde Nebelflor
Erhellte sich, gehorchend seinem Rohr,
Und liess in jenen Flammenschooss ihn sehn,
Wo dort sich Sonnen bilden und vergehn.
Er beugte, spaltete und maass den Strahl,
Verband, zerstreut' ihn nach Gesetz und Wahl.
Er hielt das Licht des Sirius gebannt,
Der Wega Schimmer spielt in seiner Hand.
Und während so sein Ruhm die Welt durchzog,
Sich immer mehrend über Meere flog;
Blieb' still er, sanft demüthig wie ein Kind,
Voll Herzenseinfalt und stets fromm gesinnt.

3. Helmholtz 1887, p. 115.

4. Ibid, p. 117.

5. Ibid.

6. Ibid. p. 120.

7. Ibid., p. 121.

8. Ibid.

9. For the full quotation, see chapter 5 above. Helmholtz translates Brewster's entire quotation into German for his speech.

10. Helmholtz 1887, pp. 122–123. (This note and the next refer to Foerster's speech.)

11. Ibid., p. 124.

12. Ibid., p. 126. The original German reads:

Rauh, wie des Bergstroms wilddurchbraustes Felsbett,
War Deine Jugend, Mühe nur und Arbeit,
Aber Dein Genius bahnte Dir die Pfade
Ewigen Ruhmes.
Fest auf der Praxis ehrnem Fundamente
Hast Du des Lichtes Theorie gegründet,
Bis in die fernsten Lande trugst den Ruf Du
Deutscher Mechanik.

13. Ibid., p. 128.

14. Ibid., p. 127.

15. Cahan 1993, p. 561. See also Ringer 1969.

16. Cahan 1989, p. 51.

17. Cahan 1993, pp. 586–599. Although Helmholtz clearly wanted a more unified Germany, he was certainly not a rabid nationalist (as were, say, Johann Gustav Droysen and Heinrich von Treitschke). Indeed, he abhorred the xenophobic tencencies of several of his colleagues.

18. Fuess 1887, p. 149.

19. Abbe 1989, volume 2, pp. 319–338.

20. Ibid., p. 320.

21. Ibid.

22. Ibid., p. 321.

23. Ibid., p. 326.

24. Ibid., pp. 326–327.

25. Ibid., p. 327.

26. Ibid, p. 331.

27. Ibid., p. 328.

28. Ibid., pp. 328–329.

29. In reality, however, the future of German instrument makers looked bleak indeed, as their status declined during the early Kaiserreich.

30. Abbe 1989, volume 2, p. 329.

31. Ibid.

32. Ibid, p. 332.

33. Ibid., p. 337.

34. Ibid., pp. 335–336.

35. Ibid., p. 336.

36. Wittig 1989, pp. 36–40.

37. Ibid., p. 15.

38. See Feffer 1994 and Auerbach 1927.

39. Abbe 1989, volume 4, pp. 1–26.

40. Ibid., p. 4.

41. Ibid., p. 5.

42. Ibid., p. 9.

43. Ibid., p. 11.

44. Ibid., p. 10.

45. Ibid., pp. 23–24.

46. Ibid., p. 23.

47. Ibid., p. 24.

48. Ibid., pp. 24–25.

49. Ibid., pp. 27–31.

50. Ibid., p. 25.
51. Ibid., pp. 21–26.
52. Auerbach 1927, p. 223.
53. Cahan 1999, p. 302.
54. Ibid., p. 303.
55. Terry Shinn, The Research-Technology Incubus: German Origins, 1870–1900 (manuscript). I thank Terry for permission to draw from this work in my argument.
56. Abbe 1989, volume 4, p. 29.
57. Ibid., pp. 32–35.
58. Ibid., p. 33.
59. Ibid., p. 34.
60. Ibid.
61. Ibid., p. 36.
62. Ibid., pp. 39–56.
63. Ibid., pp. 48–50.
64. Ibid., pp. 72–73.
65. Ibid., p. 52.
66. Ibid., p. 50.
67. Ibid., p. 51.
68. Ibid., p. 55.
69. Ibid.
70. Ibid., pp. 78–86.
71. Ibid., p. 79.
72. Ibid., p. 80.
73. Ibid.
74. Ibid.
75. Ibid., p. 81.
76. Ibid., p. 83.
77. Ibid., p. 55.
78. Ibid.
79. Ibid., pp. 58–65.
80. Ibid., pp. 60–62.
81. Ibid., p. 129.
82. Ibid., p. 92.
83. Ibid., pp. 92–93.
84. Ibid., p. 94.
85. Ibid., p. 97.

86. Ibid., p. 107.

87. Ibid., p. 122.

88. As quoted in Cahan 1989 (p. 50). Original: *Stereographische Berichte über die Verhandlungen des Reichstages* 93 (1887): 298–310.

89. Wittig 1989, p. 15.

90. Faulhaber 1904, p. 464. Of the 10,936 employees of the German optical industry, approximately two-thirds worked in the 29 large firms employing more than 150 workers.

91. Ibid.

92. Ibid., pp. 469–471.

93. Ibid., p. 470.

94. Ibid., p. 471.

95. Ibid., p. 470.

96. See, e.g., Elkana 1974, pp. 97–145.

97. On the importance of standards to the physics community, see Olesko 1991, 1995.

98. For a history of electrical standards, see Hunt 1991, pp. 53–57.

99. Cahan 1989, p. 27.

100. Ibid., p. 11; DuBois-Reymond 1912, volume 1, pp. 393–421.

101. Virchow 1871, pp. 75, 77.

102. Virchow 1866, p. 57.

103. Ibid., p. 59.

104. Ibid., p. 60.

105. Virchow 1871, pp. 73–82.

106. Cahan 1989, p. 32.

107. As quoted in ibid. (p. 33). Original: "Votum des Herrn Geheimen Regierungs-Rathes, Prof. Dr. von Helmholtz: Ueber die Aufgaben der wissenschaftlichen Abtheilung des in Aussicht genommenen Physikalisch-Mechanischen Instituts," in *Drucksachen zu den Verhandlungen des Bundesraths des Deutschen Reiches*, Aktenstück no. 50 (Berlin, 1886), pp. 33–36.

108. As quoted in Cahan 1989 (p. 38).

109. As quoted in Cahan 1989 (p. 38).

110. Utzschneider 1826.

111. Bunsen and Kirchhoff 1860; Kirchhoff 1863. See also Sutton 1976; McGucken 1969.

112. Helmholtz 1887, pp. 115, 121–122.

113. See, e.g., Rohr 1928a, pp. 438–453, 501–514, 548–559, 600–612.

114. Feffer 1994.

115. Quoted in Zschimmer 1909 (p. 70) and Feffer 1994 (p. 193).

116. Feffer 1994, p. 233.
117. As quoted in Chance 1919 (p. 182).

Chapter 9

1. I was pleased to have the support of the Optical Society of Southern California when I offered them a version of chapters 3 and 4 in a lecture. Even today, it seems, artisanal skill is necessary in the production of optical lenses, and which aspects of lensmaking can be performed by machines and which aspects cannot are very real issues.

Appendix

1. See, e.g., SPKB.Haus II.HA.SD.Fraunhofer Nachlass, Box 2, p. 426; Box 3, p. 83.

Bibliography

Abbe, Ernst. 1989. *Gesammelte Abhandlungen* (5 volumes). Hildesheim: Georg Olms.

Airy, George Biddell. 1827. "On the Principles and Construction of the Achromatic Eye Pieces of Telescopes, and on the Achromaticism of Microscopes." *Transactions of the Cambridge Philosophical Society* 2, part 2: 227–252.

Alder, Ken. 1997. *Engineering the Revolutions: Arms, Enlightenment, and the Making of Modern France, 1763–1815*. Princeton University Press.

Amann, Joseph. 1908. *Die bayerische Landesvermessung in ihrer geschichtlichen Entwicklung*, part 1, *Die Aufstellung des Landesvermessungswerkes 1808–1871*. Munich: K. B. Katasterbureau.

Amann, Joseph. 1920. *Das bayerische Kataster*. Stuttgart: K. Wittwer.

Anderson, Warwick. 1992. "The Reasoning of the Strongest: The Polemics of Skill and Science in Medical Diagnosis." *Social Studies of Science* 22: 653–684.

The Artisan; or, Mechanic's Instructor: Containing a Popular, Comprehensive, and Systematic View of the Following Sciences: . . . Optics. 1825. Volume 1. London: William Cole.

The Artizan. 1844. *A Monthly Journal of the Operative Arts* 1.

Ashworth, William J. 1996. "Memory, Efficiency, and Symbolic Analysis: Charles Babbage, John Herschel, and the Industrial Mind." *Isis* 87: 629–653.

Atzema, Eisso J. 1993. The Structure of Systems of Lines in Nineteenth-Century Geometrical Optics: Malus' Theorem and the Description of the Infinitely Thin Pencil. Ph.D. dissertation, University of Utrecht.

Auerbach, Felix. 1927. *The Zeiss Works and the Carl Zeiss Foundation in Jena*, translated by R. Kanthack. London: W. & G. Foyle.

Babbage, Charles. 1830. *Reflexions on the Decline of Science in England, and on Some of Its Causes*. London: B. Fellowes.

Babbage, Charles, and David Brewster. 1830. "The Decline of Science in England and the Patent Laws." *Quarterly Review* 43: 305–342.

Barlow, Peter. 1827a. "Rules and Principles for Determining the Dispersive Ratio of Glass; and for Computing the Radii of Curvature for Achromatic Object-Glasses,

Submitted to the Test of Experiment." *Philosophical Transactions of the Royal Society of London* 117: 231–267.

Barlow, Peter. 1827b. "A Proposition for Carrying on a Course of Experiments, with a View to Constructing, as a National Instrument, a Large Refracting Telescope, with a Fluid Concave Lens, Instead of the Usual Lens of Flint Glass. Addressed to His Royal Highness the Lord High Admiral, and the Right Honourable and Honourable Members of the Board of Longitude." *Edinburgh New Philosophical Journal* 4: 323–329.

Barlow, Peter. 1828a. "An Account of a Series of Experiments Made with a View to the Construction of an Achromatic Telescope with a Fluid Concave Lens, Instead of the Usual Lens of Flint Glass. In a Letter Addressed to Davies Gilbert, Esq. M. P. President of the Royal Society." *Philosophical Transactions of the Royal Society of London* 118: 105–112.

Barlow, Peter. 1828b. "Experiments Relative to the Effect of Temperature on the Refractive Index and Dispersive Power of Expansible Fluids, and on the Influence of These Changes in a Telescope with a Fluid Lens." *Philosophical Transactions of the Royal Society of London* 118: 313–317.

Barlow, Peter. 1829. "An Account of the Preliminary Experiments and Ultimate Construction of a Refracting Telescope of 7.8 inches Aperture, with a Fluid Concave Lens. In a Letter Addressed to Davies Gilbert, Esq. M. P. President of the Royal Society." *Philosophical Transactions of the Royal Society of London* 119: 33–46.

Barlow, Peter. 1832. "On the Performance of Fluid Refracting Telescopes, and on the Applicability of This Principle of Construction to Very Large Instruments." *Edinburgh Journal of Science,* n.s. 6: 353–360.

Bauer, Hermann. 1991. "Die Bildprogramme des 18. Jahrhunderts in bayerischen Klöster: Eine Selbstbestätigung vor dem drohenden Ende." In *Glanz und Ende der alten Klöster*, ed. J. Kirmeier and M. Treml. Munich: Haus der Bayerischen Geschichte.

Bauernfeind, Carl Maximilian von. 1885. *Johann Georg von Soldner und sein System der bayerischen Landesvermessung*. Munich: G. Franz'schen Hof-Buch- u. Kunsthandlung.

Bauernfeind, Carl Maximilian von. 1887. *Gedächnisrede auf Joseph von Fraunhofer zur Feier seines hundertsten Geburtstags*. Munich: Verlag der königlichen bayerischen Akademie der Wissenschaften.

Behagg, Clive. 1982. "Secrecy, Ritual and Folk Violence: The Opacity of the Workplace in the First Half of the Nineteenth Century." In *Popular Culture and Custom in Nineteenth-Century England*, ed. R. Storch. Croom Helm.

Bennett, J. A. 1980. "Robert Hooke as Mechanic and Natural Philosopher." *Notes and Records of the Royal Society* 35: 33–48.

Bennett, J. A. 1983. *Science at the Great Exhibition*. Cambridge: Whipple Museum of the History of Science.

Bennett, J. A. 1985. "Instrument Makers and the 'Decline of Science in England': The Effects of Institutional Change on the Elite Instrument Makers of the Early

Nineteenth Century." In *Nineteenth-Century Scientific Instruments and Their Makers,* ed. P. DeClercq. Amsterdam: Edition Rodopi.

Bennett, J. A. 1986. "The Mechanics' Philosophy and the Mechanical Philosophy." *History of Science* 24: 1–28.

Benzenberg, Johann Friedrich. 1818. *Über das Kataster*. Bonn: Eduard Weber.

Berg, Maxine. 1980. *The Machinery Question and the Making of Political Economy 1815–1848*. Cambridge University Press.

Blair, Archibald. 1827. "On the Permanency of Achromatic Telescopes Constructed with Fluid Object-Glasses." *Edinburgh Journal of Science* 7: 336–342.

Blair, Robert. 1794. "Experiments and Observations on the Inequal Refrangibility of Light." *Edinburgh Transactions* 3, part II: 3–76.

Blau, Josef. 1940. "Das geheime Rezeptenbuch des Glasmeisters Joh. Bapt. Eisner in Klostermühle 1842–1862." *Glastechnische Berichten* 18: 12–20.

Blau, Josef. 1954, 1956. *Die Glasmacher in Böhmer- und Bayernwald in Volkskunde und Kulturgeschichte* (2 volumes). Regensburg: Michael Lassleben Kallmünz.

Boegehold, Hans. 1935. "Die Leistungen von Clairaut und d'Alembert für die Theorie des Fernrohrobjektivs und die französischen Wettbewerbsversuche gegen England in den letzten Jahrzehnten des 18. Jahrhunderts." *Zeitschrift für Instrumentenkunde* 55: 97–111.

Bontemps, Georges. 1851. *Examen historique et critique des verres, vitraux, cristaux composant la classe 24 de l'Exposition universelle de 1851*. Paris: Mathias.

Bontemps, Georges. 1868. *Guide de Verrier; traité historique et pratique de la fabrication des verres, cristaux, vitraux*. Paris: Librairie du Dictionnaire des arts et manufactures.

Boscovich, P. Rogerii Josephi. 1767. *Dissertationes quinque ad dioptricam pertinentes*. Regensburg: Johannis Thomae.

Brachner, Alto, and Max Seeberger, eds. 1976. *150. Todesjahr Joseph von Fraunhofers 1787–1826. Reden und Ansprachen*, Fraunhofer-Gesellschaft zur Förderung der angewandten Forschung e.V. Ingolstadt: Courier-Druckhaus.

Brewster, David. 1813. *A Treatise on New Philosophical Instruments, for Various Purposes in the Arts and Sciences with Experiments on Light and Colours*. Edinburgh: William Blackwood; London: John Murray.

Brewster, David. 1818. "Description of a Method of Making Doubly Refracting Prisms Perfectly Achromatic." *Annals of Philosophy, or Magazine of Chemistry, Mineralogy, Mechanics, and the Arts* 11: 175–177.

Brewster, David. 1821. "History of Mechanical Inventions and Processes in the Useful Arts." *Edinburgh Philosophical Journal* 5: 113–116.

Brewster, David. 1822. "Observations on Vision through Coloured Glasses, and on Their Application to Telescopes, and to Microscopes of Great Magnitude." *Edinburgh Philosophical Journal* 6: 102–107.

Brewster, David. 1823. "Description of a Monochromatic Lamp for Microscopial Purposes, & c. with the Remarks on the Absorption of the Prismatic Rays by Coloured Media." *Transactions of the Royal Society of Edinburgh* 9: 433–444.

Brewster, David. 1824a. "Description of a Monochromatic Lamp for Microscopical and Other Purposes" (abridged version of Brewster 1823). *Edinburgh Philosophical Journal* 10: 120–125.

Brewster, David. 1824b. "Some Account of the Schools of Arts of Edinburgh." *Edinburgh Philosophical Journal* 11: 203–205.

Brewster, David. 1825a. "Description of Fraunhofer's Large Achromatic Telescope." *Edinburgh Journal of Science* 2: 305–307.

Brewster, David. 1825b. "Some Account of the Late M. Guinand, and of the Important Discovery Made by Him in the Manufacture of Flint Glass for Large Telescopes." *Edinburgh Journal of Science* 2: 348–354.

Brewster, David. 1826a. "Observations on the Superiority of Achromatic Telescopes with Fluid Object-Glasses, as Constructed by Dr. Blair." *Edinburgh Journal of Science* 4: 282–285.

Brewster, David. 1826b. "Farther Account of the Large Achromatic Refracting Telescope of Fraunhofer in the Observatory of Dorpat." *Edinburgh Journal of Science* 5: 105–111.

Brewster, David. 1827a. "Memoir of the Life of M. Le Chevalier Fraunhofer, the Celebrated Improver of the Achromatic Telescope and Member of the Academy of Sciences at Munich." *Edinburgh Journal of Science* 7: 1–11.

Brewster, David. 1827b. "Notice Respecting Professor Barlow's New Achromatic Telescopes with Fluid Object-Glasses." *Edinburgh Journal of Science* 7: 335–336.

Brewster, David. 1828. "Description of the New Fluid Telescope Recently Constructed by Messrs W. and T. Gilbert on a Plan Suggested by Peter Barlow." *Edinburgh Journal of Science* 8: 93–96.

Brewster, David. 1831a. *A Treatise on Optics*. London: Longman, Rees, Orme, Brown and Green and John Taylor.

Brewster, David. 1831b. "Remarks on Dr. Goring's Observation on the Use of Monochromatic Light with the Microscope." *Edinburgh Journal of Science*, n.s. 5: 143–148.

Brewster, David. 1832. "Optics." *Edinburgh Encyclopaedia* 14: 589–789.

Brewster, David. 1845. "The Earl of Rosse's Reflecting Telescope." *North British Review* 2: 175–212.

Brewster, David. 1855. "The Paris Exhibition and the Patent Laws." *North British Review* 24: 231–267.

Buchwald, Jed Z. 1989. *The Rise of the Wave Theory of Light: Optical Theory and Experiment in the Early Nineteenth Century*. University of Chicago Press.

Buchwald, Jed Z. 1993. "Design for Experimenting." In *World Changes*, ed. P. Horwich. MIT Press.

Bunsen, R., and G. Kirchhoff. 1860. "Chemische Analyse durch Spectral-beobachtungen." *Poggendorffs Annalen der Physik und Chemie* 161–189.

Cahan, David. 1989. *An Institute for an Empire: The Physikalisch-Technische Reichsanstalt, 1871–1918*. Cambridge University Press.

Cahan, David. 1993. "Helmholtz and the Civilizing Power of Science." In *Hermann von Helmholtz and the Foundations of Nineteenth-Century Science*, ed. D. Cahan. University of California Press.

Cahan, David. 1996. "The Zeiss Werke and the Ultramicrocope: The Creation of a Scientific Instrument in Context." *Archimedes* 1: 67–116.

Cahan, David. 1999. "Helmholtz als führender Wissenschaftler an der Preußischen Akademie der Wissenschaften." In *Die Königlich Preußische Akademie der Wissenschaften zu Berlin im Kaisserreich*, ed. Jürgen Kocka. Berlin: Akademie Verlag.

Callebaut, Werner (organizer and moderator). 1993. *Taking the Naturalistic Turn, or How Real Philosophy of Science Is Done*. University of Chicago Press.

Cantor, Geoffrey, David Gooding, and Frank A. J. L. James. 1991. *Faraday*. Macmillan.

Cassini, Jean Dominique de. 1810. *Memoires pour servir à l'histoire des sciences, et à celle de l'Observatoire royal de Paris*. Paris: Bleuet.

Cecil, W. 1827. "On an Apparatus for Grinding Telescopic Mirrors and Object Lenses." *Transactions of the Cambridge Philosophical Society* 2: 85–105.

Chadwick, Owen. 1975. *The Secularization of the European Mind in the Nineteenth Century*. Cambridge University Press.

Chance, James Frederick. 1919. *A History of Chance Brothers & Co.: Glass and Alkali Manufacturers*. London: Spottiswoode, Ballantyne.

Chance, W. H. S. 1937. "The Optical Glassworks at Benediktbeuern." *Proceedings of the Physical Society* 49: 433–443.

Chapman, Allan. 1996. *Astronomical Instruments and Their Uses*. Variorum.

Clairaut, A. C. 1756. "Sur les moyens de perfectionner les lunettes d'approche, par l'usage d'objectifs composés de plusieurs matières differémment réfringentes." *Mémoire de l'Académie des Sciences*: 380–437.

Clairaut, A. C. 1757. "Second Mémoire sur les moyens de perfectionner les lunettes d'approche, par l'usage d'objectifs composés de plusieurs matières differémment réfringentes." *Mémoire de l'Académie des Sciences*: 524–530.

Clairaut, A. C. 1762. "Troisiéme Mémoire les moyens de perfectionner les lunettes d'approche, par l'usage d'objectifs composés de plusieurs matières differémment réfringentes." *Mémoire de l'Académie des Sciences*, pp. 578–631.

Coddington, Henry. 1823. *An Elementary Treatise on Optics*. Cambridge: J. Smith.

Coddington, Henry. 1829. *A Treatise on the Reflexion and Refraction of Light, Being Part I of a System of Optics*. Cambridge: J. Smith.

Collins, H. M. 1974. "The TEA Set: Tacit Knowledge and Scientific Networks." *Science Studies* 4: 165–186.

Collins, H. M. 1990. *Artificial Experts: Social Knowledge and Intelligent Machines*. MIT Press.

Collins, H. M. 1992. *Changing Order: Replication and Introduction in Scientific Practice*. University of Chicago Press.

Cooper, William. 1835. *The Crown Glass Cutter and Glazier's Manual*. Edinburgh: Oliver and Boyd.

Coulter, Moureen. 1992. *Property in Ideas: The Patent Question in Mid-Victorian Britain*. Thomas Jefferson Press.

Daston, Lorraine. 1995. "The Moral Economy of Science." In *Constructing Knowledge in the History of Science*, ed. A. Thackray (*Osiris* 10: 1–24).

Daumas, Maurice. 1989. *Scientific Instruments of the Seventeenth and Eighteenth Centuries and Their Makers*. Portman.

Dodd, George. 1843. *Days at the Factories, or the Manufacturing Industry of Great Britain*. London: Knight.

Dodsworth, Roger. 1987. *Glass and Glassmaking*. Aylesbury: Shire.

Dollond, George, John Herschel, and William Pearson. 1826. "Report of the Committee Appointed by the Council of the Astronomical Society of London, for the Purpose of Examining the Telescope Constructed by Mr. Tulley, by Order of the Council." *Memoirs of the Astronomical Society of London* 2: 507–511. Reprinted in *Philosophical Magazine and Journal* 67 (1826): 377–382.

Dollond, John. 1753–54. "A Letter from Mr. John Dollond to James South, A.M. F.R.S. Concerning a Mistake in M. Euler's Theorem for Correcting the Aberration in the Object-glasses of Refracting Telescopes." *Philosophical Transactions of the Royal Society of London* 48: 289–291.

Dollond, John. 1757–58. "An Account of Some Experiments Concerning the Different Refrangibility of Light." *Philosophical Transactions of the Royal Society of London* 50: 733–743.

Dollond, Peter. 1765. "An Account of an Improvement Made By Mr. Peter Dollond in his New Telescope." *Philosophical Transactions of the Royal Society of London* 55: 55–56.

Douglas, R. W., and Susan Frank. 1972. *A History of Glassmaking*. Henley-on-Thames: G. T. Foulis.

DuBois-Reymond, Emil. 1912. *Reden*, ed. Estelle DuBois-Reymond, second enlarged edition (2 volumes). Leipzig.

Dyck, Walther von. 1912. *Georg von Reichenbach*. Munich: Deutsches Museum.

Eamon, William. 1994. *Science and the Secrets of Nature: Books of Secrets in Medieval and Early Modern Culture*. Princeton University Press.

Eisenlohr, C. F. M., ed. 1856. *Sammlung der Gesetze und internationalen Verträge zum Schutze des literarischen-artistischen Eigenthums in Deutschland, Frankreich und England*. Heidelberg: Bangel and Schmitt.

Elkana, Yehuda. 1974. *The Discovery of the Conservation of Energy*. Harvard University Press.

Eschapasse, Maurice. 1963. *L'Architecture Bénédictine en Europe*. Paris: Editions des Deux-Mondes.

Evans, David S. 1981. "John Frederick William Herschel." In *Dictionary of Scientific Biography* (6: 323–328).

Faraday, Michael. 1827. *Chemical Manipulation; Being Instructions to Students in Chemistry on the Methods of Performing Experiments of Demonstration or of Research, with Accuracy and Success*. London: W. Phillips.

Faraday, Michael. 1830. "On the Manufacture of Glass for Optical Purposes." *Philosophical Transactions of the Royal Society of London* 120: 1–57.

Faulhaber, C. 1904. "Die optische Industrie." In *Handbuch der Wirtschaftskunde Deutschlands*, volume 3. Leipzig: B. G. Teubner.

Feffer, Stuart Michael. 1994. Microscopes to Munitions: Ernst Abbe, Carl Zeiss, and the Transformation of Technical Optics. Ph.D. dissertation, University of California, Berkeley.

Feffer, Stuart Michael. 1996. "Ernst Abbe, Carl Zeiss, and the Transformation of Microscopial Optics." *Archimedes* 1: 23–66.

Fellöcker, Pater Sigmund. 1864. *Geschichte der Sternwarte der Benediktiner-Abtei Kremsmünster*. Linz: J. Feichtinger.

Ferguson, James. 1823. *Lectures in Select Subjects in Mechanics, Hydrostatics, Hydraulics, Pneumatics, Optics, Geography, and Astronomy: With Notes, and an Additional Volume, Containing the Most Recent Discoveries in the Arts and Sciences by David Brewster*, third edition (2 volumes). Edinburgh: Stirling Slade and Bell & Bradfute.

Ferguson, W. T., compiler. 1961. *Sir John Herschel and Education at the Cape, 1834–1840*. Oxford University Press.

Feyerabend, Paul. 1962. "Explanation, Reduction, and Empiricism." In *Scientific Explanation, Space, and Time*, ed. H. Feigl and G. Maxwell. University of Minnesota Press.

Foucault, Michel. 1979. *Discipline and Punish: The Birth of the Prison*. Vintage.

Foucault, Michel. 1980. "Questions on Geography." In *Power/Knowledge*, ed. C. Gordon. Harvester.

Friedl, Paul. 1973. *Glasmachergeschichten und Glashüttensagen aus dem Bayerischen Wald und dem Böhmerwald*. Grafenau: Morsak.

Friedrich, Conrad. 1910. *Georg Friedrich Brander und sein Werk*. München: J. G. Weiss'schen.

Fuess, Rudolf. 1887. "Vereinsnachrichten- Deutsche Gesellschaft für Mechanik und Optik." *Zeitschrift für Instrumentenkunde* 7: 149.

Galison, Peter. 1985. "Bubble Chambers and the Experimental Workplace." In *Observations, Experiment, and Hypothesis in Modern Physical Science*, ed. P. Achinstein and O. Hannaway. MIT Press.

Galison, Peter. 1987. *How Experiments End*. University of Chicago Press.

Galison, Peter. 1997. *Image and Logic: A Material Culture of Microphysics*. University of Chicago Press.

Ganzenmüller, Wilhelm. 1956. *Beiträge zur Geschichte der Technologie und der Alchemie*. Weinheim: Verlag Chemie.

Gasquet, Cardinal (translator). 1966. *The Rule of Saint Benedict*. Cooper Square.

Gerardy, Thomas. 1955. "Episoden aus der Gaußschen Triangulation des Königreichs Hannover (1821–1844)." *Zeitschrift für Vermessungswesen* 80: 54–62.

Giddens, Anthony. 1984. *The Constitution of Society: Outline of the Theory of Structuration*. Polity.

Giedion, Siegfried. 1969. *Mechanization Takes Command*. Norton.

Gieseke, Ludwig. 1957. *Die geschichtliche Entwicklung des deutschen Urheberrechts*. Göttingen: Otto Schwarz.

Gilbert, Ludwig Wilhelm. 1810a. "Bericht, abgestattet der mathemat.-physikalischen Klasse des Instituts in der Sitzung am 10. April 1809 von HH. De Prony, Guyton Morveau und Rochon über das schwere Krystallglas zu achromatischen Objectiven welches Hr. Du Fougerais, kaiser. Glas-Fabrikant dem Institut vorgelegt hat." *Annalen der Physik* 4 (new series; volume 34 of original series): 240–257.

Gilbert, Ludwig Wilhelm. 1810b. "Auszug aus einem Berichte der HH. Delambre, Charles, Burckhardt und Gay-Lussac an die erste Klasse des Instituts, über ein schweres Krystallglas, welches die Herren Kruines und Lancon dieser Klasse vorgelegt haben." *Annalen der Physik* 4 (new series; volume 34 of original series): 460–462.

Gooding, David. 1982. "Empiricism in Practice: Teleology, Economy, and Observation in Faraday's Physics." *Isis* 73: 46–67.

Gooding, David, and Frank A. J. L. James, eds. 1985. *Faraday Rediscovered: Essays on the Life and Work of Michael Faraday, 1791–1869*. Macmillan.

Greenaway, F., M. Berman, S. Forgan, and D. Chilton, eds. 1971–1976. *Archives of the Royal Institution, Manager's Meetings, 1799–1903* (15 volumes in 7). London.

Großmann, W. 1955a. "Zwei Briefe von C. F. Gauß zur Unterweisung Hannoverscher Generalstabsoffiziere." *Zeitschrift für Vermessungswesen* 80: 45–54.

Großmann, W. 1955b. "Gauß' geodätische Tätigkeit im Rahmen zeitgenössischer Arbeiten." *Zeitschrift für Vermessungswesen* 80: 371–384.

Guttery, D. R. 1956. *From Broad-Glass to Cut Crystal: A History of the Stourbridge Glass Industry*. Leonard Hill.

Hacking, Ian. 1983. *Representing and Intervening: Introductory Topics in the Philosophy of Science*. Cambridge University Press.

Haller, Reinhard. 1975. *Historische Glashütten in den Bodenmaiser Wäldern: Ein Beitrag zur Geschichte des Glases im Bayerischen Wald*. Bodenmais: Joska-Glaskunstwerkstätten.

Hannaway, Owen. 1986. "Laboratory Design and the Aim of Science: Andreas Libavius versus Tycho Brahe." *Isis* 77: 585–610.

Hanson, N. R. 1961. *Patterns of Discovery: An Inquiry into the Conceptual Foundations of Science*. Cambridge University Press.

Helmholtz, Hermann von. 1887. "Festbericht über die Gedenkfeier zur hundertjährigen Wiederkehr des Geburtstages Josef [*sic*] Fraunhofer's." *Zeitschrift für Instrumentenkunde* 7: 113–128.

Helmholtz, Hermann von. 1903. "Ueber das Verhältnis der Naturwissenschaften zur Gesammtheit der Wissenschaften." In *Vorträge und Reden*, fifth edition (2 volumes). Braunschweig: Vierweg.

Herschel, J. F. W. 1821. "On the Alterations of Compound Lenses and Object-Glasses." *Philosophical Transactions of the Royal Society of London* 111: 222–267.

Herschel, J. F. W. 1822a. "Practical Rules for the Determination of the Radii of a Double Achromatic Object-glass." *Edinburgh Philosophical Journal* 6: 361–370.

Herschel, J. F. W. 1822b. "On a Remarkable Peculiarity in the Law of the Extraordinary Refraction of Differently-Coloured Rays Exhibited by Certain Varieties of Apophyllite." *Transactions of the Cambridge Philosophical Society* 1: 241–247.

Herschel, J. F. W. 1823. "On the Absorption of Light by Coloured Media, and the Colours of the Prismatic Spectrum Exhibited by Certain Flames; with an Account of a Ready Mode of Determining the Absolute Dispersive Power of any Medium, by Direct Experiment." *Transactions of the Royal Society of Edinburgh* 9: 445–460.

Herschel, J. F. W. 1826. "Observations on Mr. Fraunhofer's Memoir on the inferiority of Reflecting Telescopes When Compared with Refractors." *Quarterly Journal of Science, Literature, and the Arts* 20: 288–293.

Herschel, J. F. W. 1830. *Preliminary Discourse on the Study of Natural Philosophy*. London: Longman, Rees, Orme, Brown and Green and John Taylor.

Hesse, Mary. 1963. *Models and Analogies in Science*. Sheed & Ward.

Hillier, Bill, and Julienne Hauser. 1984. *The Social Logic of Space*. Cambridge University Press.

Hobsbawm, Eric, and Terence Ranger, eds. 1983. *The Invention of Tradition*. Cambridge University Press.

Holton, Gerald. 1978. *The Scientific Imagination*. Cambridge University Press.

Howard, Alexander L. 1940. *The Worshipful Company of Glass-Sellers of London*. Glass-Sellers Co.

Howard-Duff, Ian. 1987. "Joseph Fraunhofer (1787–1826)." *Journal of the British Astronomical Association* 97: 339–347.

Hunt, Bruce. 1991. *The Maxwellians*. Cornell University Press.

Imhof, Maximus. 1806. "Verzeichnis der physikalischen, mathematischen und chemischen Apparates bey der Königlichen Akademie der Wissenschaften in München." (Deutsches Museum Archiv 1954/52).

Jackson, Myles W. 1992a. "Goethe's Economy of Nature and the Nature of His Economy." *Accounting, Organizations, and Society* 17: 459–469.

Jackson, Myles W. 1992b. "The Politics of Goethe's Views on Nature." *Enlightenment* 2: 143–157.

Jackson, Myles W. 1993. "Die britische Antwort auf Fraunhofer und die bayerische Hegemonie der Optik." In *Deutsches Museum Jahrbuch 1992*. Munich: C. H. Beck.

Jackson, Myles W. 1994a. "A Spectrum of Belief: Goethe's 'Republic' versus Newtonian 'Despotism,'" *Social Studies of Science* 24: 673–701.

Jackson, Myles W. 1994b. "Natural and Artificial Budgets: Accounting for Goethe's Economy of Nature." *Science in Context* 7: 409–431. Reprinted in *Accounting and Science*, ed. M. Power. Cambridge University Press.

Jackson, Myles W. 1994c. "Artisanal Knowledge and Experimental Natural Philosophers: Focusing on the British Response to Joseph von Fraunhofer's Optical Institute." *Studies in History and Philosophy of Science* 25: 549–575.

Jackson, Myles W. 1995. "Genius and the Stages of Life in Eighteenth-Century Britain and Germany." In *Les Ages de la Vie en Grande Bretagne au XVIIIe Siècle*, ed. S. Soupel. Paris: Presses de la Sorbonne Nouvelle.

Jackson, Myles W. 1996. "Buying the Dark Lines of the Solar Spectrum: Joseph von Fraunhofer and His Standard for Optical Glass Production." *Archimedes* 1: 1–22.

Jackson, Myles W. 1999a. "Illuminating the Opacity of Achromatic Lenses Production: Joseph von Fraunhofer and His Monastic Laboratory." In *The Architecture of Science*, ed. P. Galison and E. Thompson. MIT Press.

Jackson, Myles W. 1999b. "The State and Nature of Freedom: German Romantic Biology and Ethics." In *Essays in the Philosophy of Biology*, ed. M. Ruse and J. Maienschein. Cambridge University Press.

James, Frank A. J. L., ed. 1985a. *Faraday Rediscovered: Essays on the Life and Work of Michael Faraday, 1791–1869*. Macmillan.

James, Frank A. J. L. 1985b. "The Creation of a Victorian Myth: The Historiography of Spectroscopy." *History of Science* 23: 1–24.

James, Frank A. J. L. 1991a. "Time, Tide, and Michael Faraday." *History Today*: 22–34.

James, Frank A. J. L. 1991b. "The Military Context of Chemistry: The Case of Michael Faraday." *Bulletin of the History of Chemistry* 11: 36–40.

James, Frank A. J. L., ed. 1991–1997. *The Correspondence of Michael Faraday* (3 volumes). Peregrinus.

James, Frank A. J. L. 1992. "Michael Faraday: The City Philosophical Society and the Society of Arts." *Royal Society of Arts Journal* 140: 192–199.

Jardine, Nicholas. 1991. *The Scenes of Inquiry: On the Reality of Questions in the Sciences*. Clarendon.

Jebsen-Marwedel, Hans. 1987. *Joseph von Fraunhofer und die Glashütte in Benediktbeuern*. Munich: Fraunhofer-Gesellschaft zur Förderung der angewandten Forschung.

Jörg, Leonhard. 1859. Fraunhofer und seine Verdienste um die Optik. Dissertation, Munich.

Jolly, Philipp. 1865. *Das Leben Fraunhofers*. Munich: J. G. Cottascher Verlag.

Jones, Edmund D. 1975. *English Critical Essays: Sixteenth, Seventeenth and Eighteenth Centuries.* Oxford University Press.

Kant, Immanuel. 1988. *The Critique of Judgement.* Clarendon.

King, Henry C. 1955. *The History of the Telescope.* Sky.

Kirchhoff, G. R. 1863. "Zur Geschichte der Spectral-Analyse und der Analyse der Sonnenatmosphäre." *Poggendorffs Annalen der Physik und Chemie* 118: 94–111.

Kirmeier, Josef, and Manfred Treml, eds. 1991. *Glanz und Ende der alten Klöster: Säkularisation im bayerischen Oberland 1803.* Munich: Haus der Bayerischen Geschichte.

Klingenstierna, Samuel. 1759–60. "De Aberratione Luminis, in Superficiebus et Lentibus Sphaericis refractorum" *in Philosophical Transactions of the Royal Society of London* 51: 944–977.

Klügel, Georg Simon. 1778. *Analytische Dioptrik in zwey Theilen: Der erste enthällt die allgemeine Theorie der optischen Werkzeuge; der zweyte die besondere Theorie und vortheilhafteste Einrichtung aller Gattungen von Fernröhren, Spiegelteleskopen, und Mikroskopen.* Leipzig: Johann Friedrich Junius.

Klügel, Georg Simon. 1810a. "Angabe eines möglichst vollkommenen achromatischen Doppel-Objectivs, und über die Anwendbarkeit dieser und ähnlicher Berechnungen für Künstler zur Verfertigung achromatischer Fernröhre." *Annalen der Physik* 4 (new series; volume 34 of original series): 265–275.

Klügel, Georg Simon. 1810b. "Weitere Entwicklung der Angabe eines vollkommenen Doppel-Objectivs in dem vorherliegenden Aufsatze." *Annalen der Physik* 4 (new series; volume 34 of original series): 276–291.

Koch, Ernst-Eckhard. N.d. "Das Konservatorenamt und die mathematisch-physikalische Sammlung der Bayerischen Akademie der Wissenschaft" (manuscript in Deutsches Museum Library, Munich, Signatur: SB 2150).

Koch, H. 1936a. "Neues über den Hofmechanikus Dr. Friedrich Körner." *Zeiss-Werkzeitung* 11: 60–64.

Koch, H. 1936b. "Die jenaischen Universitäts-Mechanici." *Zeiss-Werkzeitung* 11: 34–37.

Koch, H. 1945. "Johann Friedrich Braunau: der Amtsvorgänger von Carl Zeiss als Universitätsmechaniker." *Zeiss-Werkzeitung* 18: 9–10.

Körner, Friedrich. 1831. *Anleitung zur Bearbeitung des Glases an der Lampe, und zur vollständigen Verfertigung der, durch das Lampenfeuer darstellbaren, physikalischen und chemischen Instrumente und Apparate.* Jena: August Schmid.

Kuhn, Thomas S. 1962. *Structures of Scientific Revolutions.* University of Chicago Press.

Kuhn, Thomas S. 1977. *The Essential Tension: Selected Studies in Scientific Tradition and Change.* University of Chicago Press.

Latour, Bruno. 1983. "Give Me a Laboratory and I Will Raise the World." In *Science Observed*, ed. K. Knorr-Cetina and M. Mulkay. Sage.

Latour, Bruno. 1990. "*Post*modern? No, Simply *A*modern: Towards an Anthropology of Science." *Studies in History and Philosophy of Science* 21: 145–171.

Lawrence, Chris. 1985. "Incommunicable Knowledge: Science, Technology, and the Clinical Art in Britain 1850–1914." *Journal of Contemporary History* 20: 503–520.

Lectures on the Results of the Great Exhibition of 1851, Delivered before the Society of Arts, Manufactures, and Commerces, at the Suggestion of H. R. H. Prince Albert, President of the Society. 1852. London.

Lefebvre, Henri. 1984. *The Production of Space*. Blackwell.

Lerner, Franz. 1981. *Geschichte des deutschen Glaserhandwerks*, second edition. Schorndorf: Hofmann-Verlag.

Littrow, J. J. 1831. "On Barlow's New Telescopes." *Memoirs of the Astronomical Society of London* 4: 481–493.

Lobmeyr, L. 1874. *Die Glasindustrie, ihre Geschichte, gegenwärtige Entwicklung und Statistik*. Stuttgart: W. Spemann.

Lommel, E., ed. 1888. *Joseph von Fraunhofer's Gesammelte Schriften: Im Auftrage der Mathematisch-Physikalischen Classe der königlichen bayerischen Akademie der Wissenschaften*. Munich: Verlag der königlichen Akademie in Commission bei G. Franz.

Long, Pamela O. 1991. "Invention, Authorship, 'Intellectual Property,' and the Origin of Patents: Notes toward a Conceptual History." *Technology and Culture* 32: 864–884.

MacLeod, Christine. 1996. "Concepts of Invention and the Patent Controversy in Victorian Britain." In *Technological Change*, ed. R. Fox. Harwood.

McGucken, W. 1969. *Nineteenth-Century Spectroscopy: Development of the Understanding of Spectra, 1802–1897*. Baltimore: Johns Hopkins University Press.

MacKenthun, I. 1958. *Joseph von Utzschneider, sein Leben, sein Wirken, seine Zeit*. Munich: Universitäts-Druck.

Matsumura, Takao. 1983. *The Labour Aristocracy Revisited: The Victorian Flint Glass Makers*. Manchester, UK: Manchester University Press.

The Mechanics Weekly Journal; or Artisan's Miscellany of Inventions, Experiments, Projects, and Improvements in the Useful Arts. 1824. London: Westley and Parrish.

Merrifield, Mary P., ed. 1849. *Original Treatises Dating from the XII to the XVIII Centuries, on the Arts of Painting* (2 volumes). London: J. Murray.

Merz, Sigmund. 1865. *Das Leben und Wirken Fraunhofers*. Landshut: Joseph Thomann'schen Buchhandlung.

Messerschmidt, E. 1976. "Die wissenschaftlichen Grundlagen der bayerischen Landesvermessung." In *Das öffentliche Vermessungswesen in Bayern*. Munich: Bayerisches Staatsministerium der Finanzen und Vermessungssverwaltung.

Morrell, J. B. 1974. "Science in Manchester at the University of Edinburgh." In *Artisan to Graduate*, ed. D. Cardwell. Manchester University Press.

Morrell, Jack, and Arnold Thackray. 1981. *Gentlemen of Science: Early Years of the British Association for the Advancement of Science*. Clarendon.

Morrison-Low, A.D. 1984. "Brewster and Scientific Instruments." In *"Martyr of Science": Sir David Brewster, 1781–1868*, ed. A. Morrison Low and J. Christie. Edinburgh: Royal Society Museum.

Morse, Edgar. 1981. "David Brewster." In *Dictionary of Scientific Biography*, volume 2. Scribner.

Morus, Iwan Rhys. 1998. *Frankenstein's Children: Electricity, Exhibition, and Experiment in Early-Nineteenth-Century London*. Princeton University Press.

Mumford, Lewis. 1967. *The Myth of the Machine*, volume 1: *Technics and Human Development*. Secker and Warburg.

Newman, William. 1999. "Alchemical Symbolism and Concealment: The Chemical House of Libavius." In *The Architecture of Science*, ed. P. Galison and E. Thompson. MIT Press.

Newton, Isaac. 1952. *Opticks, or a Treatise of the Reflections, Refractions, Inflections & Colours of Light*. Dover.

Nickles, Thomas Jacob. 1980a. *Scientific Discovery, Logic, and Rationality*. Reidel.

Nickles, Thomas Jacob. 1980b. *Scientific Discovery: Case Studies*. Reidel.

Niedersächsische Vermessungs- und Katasterverwaltung, ed. 1955. *C. F. Gauss und die Landesvermessung in Niedersachsen*. Hannover: Niedersächisches Landesvermessungsamt.

Nipperdey, Thomas. 1991. *Deutsche Geschichte, 1800–1866, Bürgerwelt und starker Staat*. Munich: C. H. Beck.

Olesko, Kathryn. 1991. *Physics as a Calling: Discipline and Practice in the Königsberg Seminar for Physics*. Cornell University Press.

Olesko, Kathryn. 1993. "Tacit Knowledge and School Formation." *Osiris* 8: 16–29.

Olesko, Kathryn. 1995. "The Meaning of Precision: The Exact Sensibility in Early-Nineteenth-Century Germany." In *The Values of Precision*, ed. M. Norton Wise. Princeton University Press.

Olesko, Kathryn. 1996. "Precision, Tolerance, and Consensus: Local Cultures in German and British Resistance Standards." *Archimedes* 1: 117–156.

Ophir, Adi. 1984. The City and the Space of Discourse: Plato's Republic: Textual Acts and Their Political Significance. Ph.D. dissertation, Boston University.

Ovitt, Jr., George. 1986. *The Restoration of Perfection: Labor and Technology in Medieval Culture*. New Brunswick: Rutgers University Press.

Owens, Larry. 1985. "Pure and Social Government: Laboratories, Playing Fields, and Gymnasia in the Nineteenth-Century Search for Order." *Isis* 76: 182–194.

Pannabeck, John. 1998. "Representing Mechanical Arts in Diderot's *Encyclopèdie*." *Technology and Culture* 39: 33–73.

Pape, Helmut. 1970. "Klopstocks Autorenhonorare und Selbstverlagsgewinne." *Archiv für Geschichte des Buchwesens* 10: columns 1–268.

Pellatt, Apsley, Jr. 1821. *Memoir on the Origin, Progress, and Improvement of Glass Manufactures*. London: B. J. Holdsworth.

Pellatt, Apsley, Jr. 1849. *Curiosities of Glass Making: With Details of the Processes and Productions of Ancient and Modern Ornamental Glass Manufacture*. London: David Bogue.

Pickering, Andrew, ed. 1991. *Science as Culture and Practice*. University of Chicago Press.

Pickering, Andrew. 1995. *The Mangle of Practice: Time, Agency, and Science*. University of Chicago Press.

Pinch, Trevor. 1985. "Towards an Analysis of Scientific Observation: The Externality and Evidential Significance of Observational Reports in Physics." *Social Studies of Science* 15: 3–36.

Pinch, Trevor, Harry M. Collins, and Larry Carbone. 1996. "Inside Knowledge: Second Order Measures of Skill." *Sociological Review* 44: 163–186.

Polanyi, Michael. 1958. *Personal Knowledge: Towards a Post-Critical Philosophy*. Routledge and Kegan Paul.

Polanyi, Michael. 1962. "The Republic of Science: Its Political and Economic Theory." *Minerva* 1: 54–73.

Polanyi, Michael. 1967. *The Tacit Dimension*. Anchor.

Popper, Karl. 1959. *The Logic of Scientific Discovery*. Harper and Row.

Porter, G. R. 1832. "A Treatise on the Origin, Progressive Improvement, and Present State of the Manufacture of Porcelain and Glass." In *The Cabinet Cyclopaedia*, ed. D. Lardner. London: Longman, Rees, Orme, Brown and Green and John Taylor.

Powell, Harry J. 1883. *The Principles of Glass-Making*. London: George Bells.

Powell, Harry J. 1923. *Glass-Making in England*. Cambridge University Press.

Prechtl, Johann Joseph. 1828. *Praktische Dioptrik als vollständige und gemeinfaßliche Anleitung zur Verfertigung achromatischer Fernröhre: Nach den neuesten Verbessungen und Hülfsmitteln und eignen Erfahrungen*. Vienna: J. G. Heubner.

Preyß, Carl R. 1989. *Joseph von Fraunhofer. Optiker-Erfinder-Pionier*. Weilheim: Stöppel.

Prothero, Iowerth J. 1979. *Artisans and Politics in Early Nineteenth-Century London: John Gast and His Times*. Folkstone: William Dawson.

Ravetz, Jerome R. 1996. *Scientific Knowledge and Its Social Problems*. Transaction.

Reich, Philipp Erasmus. 1773. *Zufällige Gedanken eines Buchhändlers über Herrn Klopstocks Anzeige einer gelehrten Republik*. Leipzig.

Reichenbach, Georg. 1821. "Berichtigung der von Herrn Mechanikus Jos. Liebherr in München abgegebener Erklärung über die Erfindung meiner Kreiseinteilungsmethode." *Gilberts Annalen der Physik* 68: 33–59.

Reichenbach, Hans. 1938. *Experience and Prediction*. University of Chicago Press.

R[eynier], E. 1824. "Notice sur feu M. Guinand, opticien; demeurant aux Brenets, Canton de Neuchatel." *Bibliothèque Universelle, Sciences et Arts* 24: 142–158, 227–236.

Riekher, Rolf. 1990. *Fernrohre und ihre Meister*. Zweite stark bearbeitet Auflage. Berlin: Verlag Technik.

Ringer, Fritz K. 1969. *The Decline of the German Mandarins: The German Academic Community, 1890–1933*. Hanover: Wesleyan University Press.

Rogers, Alexander. 1829. "On the Construction of Large Achromatic Telescopes: Communicated to J. F. W. Herschel, Esq., President of the Astronomical Society." *Memoirs of the Astronomical Society of London* 3: 229–233.

Rohr, Moritz von. 1924. "Die Entwicklung der Kunst, optisches Glas zu schmelzen." *Die Naturwissenschaften* 12: 781–797.

Rohr, Moritz von. 1926. "Pierre Louis Guinand." *Zeitschrift für Instrumentenkunde* 46: 121–137, 189–197.

Rohr, Moritz von. 1928a. "P. L. Guinands Anweisung zum Glasschmelzen." *Zeitschrift für Instrumentenkunde* 48.

Rohr, Moritz von. 1928b. "Eine Erinnerung an Joseph Fraunhofer." *Forschungen zur Geschichte der Optik* 1: 2–6.

Rohr, Moritz von. 1929a. *Joseph Fraunhofers Leben, Leistungen und Wirksamkeit*. Leipzig: Akademische Verlagsgesellschaft.

Rohr, Moritz von. 1929b. *Forschungen zur Geschichte der Optik* 1: 42–52.

Rohr, Moritz von. 1933. "Optisches Glas rund ein halbes Jahrhundert nach Joseph Fraunhofers Tode, 1826–1876." *Zeitschrift für Instrumentenkunde* 53: 413–423, 457–465, 494–502.

Roth, Günther D. 1976. *Joseph von Fraunhofer: Handwerker-Forscher-Akademiemitglied, 1787–1826*. Stuttgart: Wissenschaftliche Verlagsgesellschaft.

Rule, John. 1987. "The Property of Skill in the Period of Manufacture." In *The Historical Meaning of Work*, ed. P. Joyce. Cambridge University Press.

Sang, Hans-Peter. 1987. *Joseph von Fraunhofer: Forscher, Erfinder, Unternehmer*. Munich: Dr. Peter Glas.

Schaffer, Simon. 1988. "Astronomers Mark Time." *Science in Context* 2: 115–145.

Schaffer, Simon. 1989. "Glass Works: Newton's Prisms and the Uses of Experiment." In *The Uses of Experiment*, ed. D. Gooding et al. Cambridge University Press.

Schaffer, Simon. 1990. "Newtonianism." In *Companion to the History of Modern Science*, ed. R. Olby et al. Routledge Kegan Paul.

Schaffer, Simon. 1991. "Genius in Romantic Natural Philosophy." In *Romanticism and the Sciences*, ed. A. Cunningham and N. Jardine. Cambridge University Press.

Schaffer, Simon. 1992. "A Manufactory of Ohms: The Integrity of Victorian Values." In *Invisible Connections*, ed. R. Bud and S. Cozzens. SPIE Press.

Schaffer, Simon. 1994. "Babbage's Intelligence: Calculating Engines and the Factory System." *Critical Inquiry* 21: 203–227.

Schaffer, Simon. 1997. "Experimenters' Techniques, Dyers' Hands, and the Electric Planetarium." *Isis* 88: 456–483.

Schaffer, Simon. n.d. Enlightened Automata. Manuscript.

Scherer, A. N. 1799. *Kurze Darstellung der chemischen Untersuchungen der Gasarten: Für seine öffentliche Vorlesungen entworfen*. Weimar: Gädicke.

Schmitz, E.-H. 1982. *Von Newton bis Fraunhofer*. Bonn: Verlag J. P. Wayenborgh.

Secord, Anne. 1994a. "Science in the Pub: Artisan Botanists in Early Nineteenth-century Lancashire." *History of Science* 32: 269–315.

Secord, Anne. 1994b. "Corresponding Interests: Artisans and Gentlemen in Natural History Exchange Networks." *British Journal for the History of Science* 27: 383–408.

Seitz, Adolf. 1923. "Der Münchner Optiker Joseph Niggl." *Central-Zeitung für Optik und Mechanik* 44: 150–154.

Seitz, Adolf. 1926. *Joseph Fraunhofer und sein Optisches Institut*. Berlin: Julius Springer.

Seitz, Adolf. 1927a. "Eine kleine Ergänzung zu der Lebensgeschichte des Münchner Optikers Josef Niggl." *Central-Zeitung für Optik und Mechanik* 48: 134–135.

Seitz, Adolf. 1927b. "Der Betrieb in der Benediktbeurer Glashütte für optisches Glas." *Deutsche Optische Wochenschrift* 13: 313–315, 318–319, 356–359.

Seitz, Adolf. 1929. "Die Utzschneider-Fraunhofersche Optische Werkstätte nach Fraunhofers Tode und das Leben Georg Merzs, des ersten Besitzers der Anstalt Utzschneiders." *Deutsche Optische Wochenschrift* 15: 55–57.

Shapin, Steven. 1984. "Brewster and the Edinburgh Career in Science." In *"Martyr of Science": Sir David Brewster, 1781–1868*, ed. A. Morrison Low and J. Christie. Edinburgh: Royal Society Museum.

Shapin, Steven. 1988. "The House of Experiment in Seventeenth-Century England." *Isis* 79: 373–404.

Shapin, Steven. 1994. *A Social History of Truth: Civility and Science in Seventeenth-Century England*. University of Chicago Press.

Shapin, Steven, and Simon Schaffer. 1985. *Leviathan and the Air-Pump: Hobbes, Boyle, and the Experimental Life*. Princeton University Press.

Shapiro, Alan E. 1993. *Fits, Passions, and Paroxysms: Physics, method, and chemistry, and Newton's theories of colored bodies and fits of easy reflection*. Cambridge University Press.

Shapiro, Alan E. 1996. "The Gradual Acceptance of Newton's Theory of Light and Color, 1672–1727." *Perspectives on Science* 4: 59–140.

Sheehan, James J. 1989. *German History, 1770–1866*. Clarendon.

Shelby, Lou R. 1976. "The 'Secret' of the Medieval Masons." In *On Pre-Modern Technology and Science*, ed. B. Hall and D. Ornet. Undena.

Shinn, Terry. n.d. The Research-Technology Incubus: German Origins, 1870–1900. Manuscript.

Sibum, H. Otto. 1995. "Reworking the Mechanical Value of Heat: Instruments of Precision and Gestures of Accuracy in Early Victorian England." *Studies in History and Philosophy of Science* 26: 73–106.

Siemens, Werner von. 1893. *Lebenserinnerungen.* Third unchanged edition. Berlin: Julius Springer.

Simms, William. 1852. *The Achromatic Telescope and Its Various Mountings, Especially the Equatorial; To Which Are Added Some Hints on Private Observations.* London: Troughton and Simms.

Smith, Adam. 1976. *An Inquiry into the Nature and Causes of the Wealth of Nations* (2 volumes). Oxford University Press.

Smith, Pamela H. 1994. *The Business of Alchemy: Science and Culture in the Holy Roman Empire.* Princeton University Press.

Smith, Robert. 1738. *A Compleat System of Opticks in Four Books, viz. a Popular, a Mathematical, a Mechanical, and a Philosophical Treatise.* Cambridge: Cornelius Crownfield.

Smith, Robert. 1778. *The Elementary Parts of Dr. Smith's Compleat System of Opticks: Selected and Arranged for the Use of Students at the Universities; To Which Are Added in the Form of Notes Some Explanatory Propositions from Other Authors.* Cambridge: J. Archdeacon.

Soldner, Johann Georg von. 1911. *Theorie der Landesvermessung*, repr., ed. J. Frischauf, Ostwald Klassiker, ser. no. 184. Leipzig: Wilhelm Engelmann.

Sorrenson, Richard. 1993. Scientific Instrument Makers at the Royal Society of London, 1720–1780. Ph.D. dissertation, Princeton University.

South, James. 1826. "Examination of the Large Achromatic of the Royal Observatory at Paris." *Quarterly Journal of Science, Literature, and the Arts* 20: 286–288.

Stahl, Wolfgang. 1929. Joseph von Utzschneider und seine Bedeutung für die deutsche Industrie. D. phil. dissertation, Salesianischen Offizin, Munich.

Star, Susan Leigh. 1991. "The Sociology of the Invisible: The Primacy of Work in the Writings of Anselm Strauss." In *Social Organization and Social Process*, ed. D. Maines. Aldine de Gruyter.

Stewart, Larry. 1998. "A Meaning for Machines: Modernity, Utility, and the Eighteenth-Century British Public." *Journal of Modern History* 70: 259–294.

Struve, F. G. W. 1825. *Beschreibung des auf der Sternwarte in Dorpat befindlichen grossen Refractors von Fraunhofer.* Dorpat: J. C. Schünmann.

Struve, F. G. W. 1826a. "An Account of the Arrival and Erection of Fraunhofer's Large Refracting Telescope at the Observatory of the Imperial University of Dorpat: Communicated in a Letter from Professor Struve to Francis Baily, Esq. President of This Society." *Memoirs of the Astronomical Society of London* 2: 93–100.

Struve, F. G. W. 1826b. "A Comparison of Observations Made on Double-Stars: Communicated in a Letter to J. F. W. Herschel, Esq., Foreign Secretary of This Society." *Memoirs of the Astronomical Society of London* 2: 443–455.

Struve, F. G. W. 1828. "Report on Double Stars, from a Review of the Starry Heavens Made with the Great Achromatic Telescope of Fraunhofer, Addressed to Prince Lieven, Curator of the University of Dorpat." *Edinburgh Journal of Science* 9: 79–90.

Stutzer, Dietmar. 1986. *Klöster als Arbeitgeber um 1800: Die bayerischen Klöster als Unternehmenseinheiten und ihre Sozialsysteme zur Zeit der Säkularisation 1803.* Göttingen: Vanderhoeck & Ruprecht.

Sutton, M. A. 1972. Spectroscopy and the Structure of Matter: A Study in the Development of Physical Chemistry. D. Phil. dissertation, Oxford University.

Sutton, M. A. 1976. "Spectroscopy and the Chemists: A Neglected Opportunity." *Ambix* 23: 16–26.

Swimme, R. 1924. "Die Anfänge der optischen Glasschmelzkunst." *Keramische Rundschau* 32: 259–260.

Thirteenth Report of the Commissioners of Inquiry into the Excise Establishment, and into the Management and Collection of the Excise Revenue throughout the United Kingdom: Glass; Presented by His Majesty's Council. 1835. London: William Clowes.

Thompson, E. P. 1963. *On the Making of the English Working Class.* Pantheon.

Utzschneider, Joseph von. 1826. *Kurzer Umriß der Lebens-Geschichte des Herrn Dr. Joseph von Fraunhofer, königlich bayerischen Professors und Akademikers, Ritters des königlich bayerischen Civil-Verdienst, und des königlich dänischen Dannebrog-Ordens, Mitgliedes mehrerer gelehrten Gesellschaften.* Munich: Rösl'schen Schriften.

Utzschneider, Joseph von. 1829. "Von Utzschneider's Erklärung der Guinand'schen Glasfabrikation." *Poggendorffs Annalen der Physik und Chemie* 15: 245–251.

Utzschneider, Joseph von. 1830. "On M. Guinand's Glass for Telescopes." *Philosophical Magazine or Annals of Chemistry, Mathematics, Astronomy, Natural History, and General Science* 7: 351–353. Originally published in *Astronomische Nachrichten* (1829), no. 163.

Utzschneider, Joseph von. 1837. *Mit welchen Schwierigkeiten begann im Jahre 1799 und 1800 die Regierung Sr. Majestät des Königs Maximilian Joseph in Bayern?* Munich: Franz Seraph Hübschmann.

Veit, H. 1962. "P. Ulrich Schiegg von Ottobeuern (1752–1810) und die bayerische Landesvermessung." *Beiträge zur Geschichte der Abtei Ottobeuern* 73: 153–171.

Virchow, Rudolf. 1866. "Ueber die nationale Entwicklung und Bedeutung der Naturwissenschaften." In *Amtlicher Bericht über die vierzigste Versammlung Deutscher Naturforscher und Ärzte zu Hannover im September 1865*, ed. C. Krause and K. Karmarsch. Hannover: Hahn'sche Hofbuchhandlung.

Virchow, Rudolf. 1871. "Über die Aufgaben der Naturwissenschaften in dem neuen nationalen Leben Deutschlands." In *Tageblatt der 44. Versammlung Deutscher Naturforscher und Aerzte in Rostock 1871* (5: 73–81).

Vopelius, Eduard. 1895. *Entwicklungsgeschichte der Glasindustrie Bayerns bis 1806.* Stuttgart: J. G. Cotta'schen Buchhandlung.

Walch, A. 1935. Die wirtschaftspolitische Entwicklung in Bayern unter Montgelas, 1799–1817. D. phil. dissertation, Universität Erlangen.

Weber, Leo. 1991. "Zur Geschichte." In *Glanz und Ende der alten Klöster*, ed. J. Kirmeier and M. Treml. Munich: Haus der Bayerischen Geschichte.

Weber, Max. 1976. *The Protestant Ethic and the Spirit of Capitalism*. Allen & Unwin.

Weis, Eberhard. 1971. *Montgelas, 1759–1799: Zwischen Revolution und Reform*. Munich: C. H. Beck.

Weis, Eberhard. 1983a. *Die Säkularisation der bayerischen Klöster 1802/03: Neue Forschungen zur Vorgeschichte und Ergebnisse*. Munich: Sitzungsberichte der Bayerischen Akademie der Wissenschaften.

Weis, Eberhard. 1983b. "Bayern und Frankreich in der Zeit des Konsulats und des Ersten Empire (1799–1815)." *Historische Zeitschrift* 237: 559–595.

Weis, Eberhard. 1991. "Die politischen Rahmenbedingungen." In *Glanz und Ende der alten Klöster*, ed. J. Kirmeier and M. Treml. Munich: Haus der Bayerischen Geschichte.

White, Lynn, Jr. 1971. "Cultural Climates and Technological Advances in the Middle Ages." *Viator* 2 : 171–201. Reprinted in L. White, Jr., *Medieval Religions and Technology*. University of California Press, 1978.

Williams, L. Pearce. 1981. "Michael Faraday." In *Dictionary of Scientific Biography* (4: 527–540).

Wittgenstein, Ludwig. 1953. *Philosophical Investigations*. Macmillan.

Wittig, Joachim. 1981. "Friedrich Körner und die Anfänge des wissenschaftlichen Gerätebaues in Jena in der ersten Hälfte des 19. Jahrhunderts." *Schriftenreihe für Geschichte der Naturwissenschaften, Technik und Medizin* 18: 17–28.

Wittig, Joachim. 1989. *Ernst Abbe*. Leipzig: G. Teubner.

Wollaston, William Hyde. 1802. "A Method of Examining Refractive and Dispersive Powers by Prismatic Reflection." *Philosophical Transactions of the Royal Society of London* 92: 365–380.

Woodmansee, Martha. 1994. *The Author, Art, and the Market: Rereading the History of Aesthetics*. Columbia University Press.

Workman, Herbert B. 1913. *The Evolution of the Monastic Ideal from the Earliest Times Down to the Coming of the Friars*. London: Charles H. Kelley.

Yelin, Ritter Julius von. 1818. "Rede gehalten in der Versammlung der Verwandlung des Verwaltungs-Ausschußes des polytechnischen Vereins in Bayern zu München am 26. August 1818 zur Feyer des Stiftungsfestes, der neuen Beamtenwohl und der Niederlegung seiner Stelle als diesjährigen Vorstands des Vereins." *Kunst- und Gewerb-Blatt des polytechnischen Vereins im König-Reiche Bayern* 4: 601–608.

Ziggelaar, August. 1993. "Placing Some of Boscovich's Contributions to Optics in the History of Physics." In *R. J. Boscovich: Vita e Attività Scientifica*. Rome: Istituto della Enciclopedia Italiana.

Zschimmer, E. 1909. *Die Glasindustrie in Jena: Ein Werk von Schott and Abbe*. Jena: E. Diedrichs.

Zschokke, Heinrich. 1817. "Die Werkstätten in Benediktbeuern." In *Ueberlieferungen zur Geschichte unserer Zeit*. Aarau.

Index

Printed in the United States
By Bookmasters